INCOMPLETE TILINGS

BEING THE SECOND PART
OF SEVERAL COMPRISING
THE COMPLETE? POLYHEDRA
BY PATRICK TAYLOR

INCOMPLETE TILINGS

First published 1998 by

Nattygrafix
277 Cavendish Street
Ipswich
Suffolk
IP3 8BQ

(01473) 289861

Made in England

ISBN 0 9516701 3 1

NATTYGRAFIX IPSWICH

CONTENTS

ILLUSTRATIONS

For Alan Monks who laid the foundations

"... we are rejecting the current fashion that geometry must be abstract if it is to be regarded as advanced mathematics, and that dispenses entirely with diagrams."

B Grünbaum & G C Shephard *Tilings and Patterns: An Introduction* 1989

INTRODUCTION

This book builds on the work already done in 'The Complete? Polygon' and looks at the ways in which polygons can be arranged in the plane. The title 'Incomplete Tilings' is a reflection of both the stature of Grünbaum & Shephard's work and the fact that the present work deals only with regular assemblies of regular and star polygons, in as much as the tilings in two dimensions mirror the possible assemblies of polygons into uniform polyhedra in three dimensions.

Whilst the polyhedra wrap around to enclose a portion of 3-D space, the tilings, filling the plane, effectively stretch to infinity cutting 3-D space in half.

With the wide variety of possible regular and star polygons already seen, there would seem to be many different ways in which these could be arranged to fill the 360° of a tiling vertex. Indeed, Grünbaum & Shephard identify 21 types of possible vertex using only regular polygons, which is further limited to just 11 when the arrangement of polygons is kept the same at each and every vertex. 10 of these comprise the three regular tilings together with their derived semi-regular forms, which have regular faces, but of more than one type.

We will herein venture beyond this limitation and allow the inclusion of other polygon types, as were described in 'The Complete? Polygon', but we will maintain the requirement that all vertices in a tiling have the same configuration. This will still exclude for example the fascinating 5-symmetric Penrose tilings and various other tilings, some of which were examined by Kepler, with star like faces that are actually concave polygons rather than true star polygons. The rewards for adopting this expanded, but still limited, outlook will hopefully become clearer as we proceed.

REGULAR TILINGS

A regular tiling is a figure that fills the plane composed of equal regular polygons equally arranged around identical vertices.

There are but three possible arrangements for such a regular tiling which use hexagons, squares and triangles respectively. These three layouts arise as a result of the 360° at each vertex being filled in turn by the internal angles of the identical polygons around it, i.e. the 3x120° of the hexagon, the 4x90° of the square or the 6x60° of the triangle.

The resultant tilings are shown in Figure 1, along with their respective polygonal vertex figures and Schläfli notation, where {x,y} indicates that y polygons of x sides each occur around each vertex.

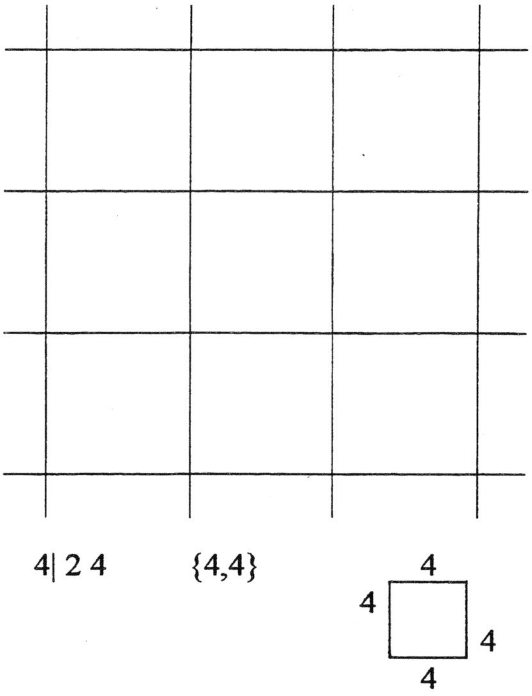

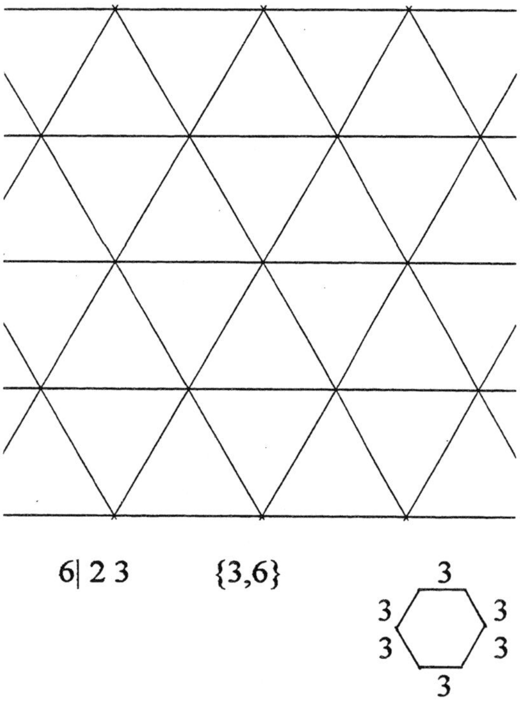

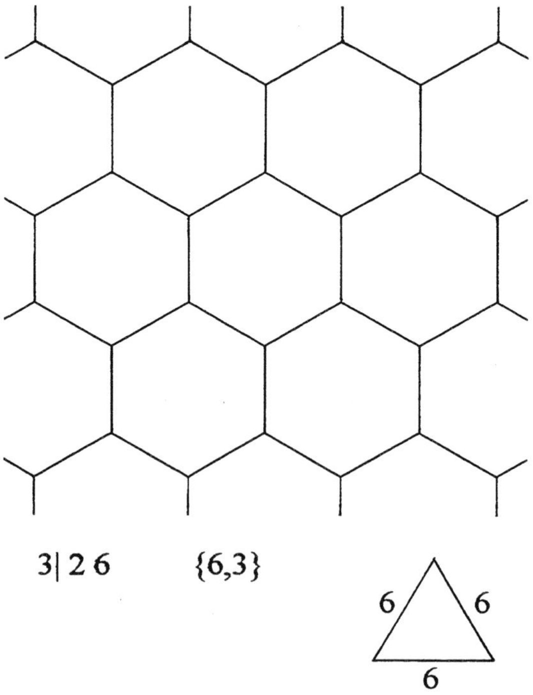

Fig. 1: Regular Tilings

DUAL TILINGS

In a similar way that the joining of the centres of edges of polygons yielded dual polygons (edge-vertex duality), we can join the centres of the faces of the polygons in a tiling to yield a dual tiling.

This process of face-vertex duality is shown in Figure 2, where each tiling can be seen to produce a dual which is in fact again one of the original tilings. Faces become vertices and vice versa so that effectively we have reversed the face and vertex symmetries to form the tiling {y,x}.

Our three regular tilings can thus be seen to comprise a dual pair of tilings, {3,6} and {6,3}, together with a self dual tiling, {4,4}.

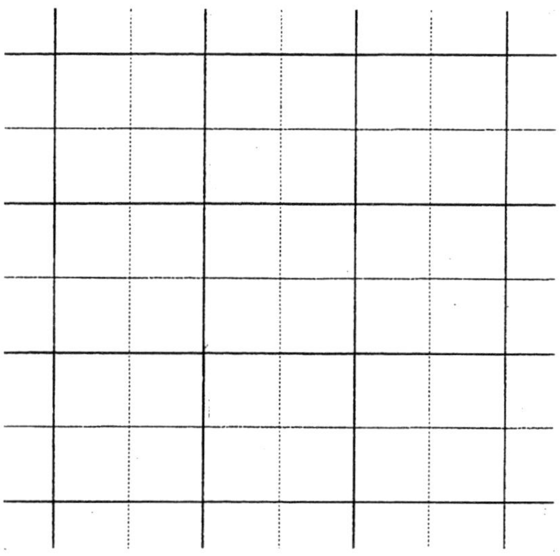

{4,4}, dual of {4,4}

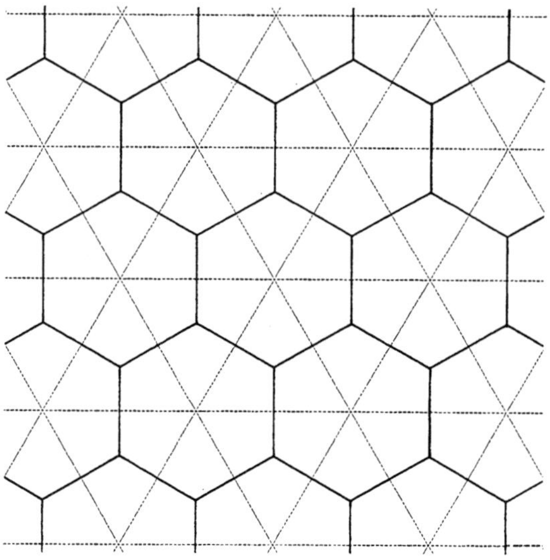

{6,3}, dual of {3,6}

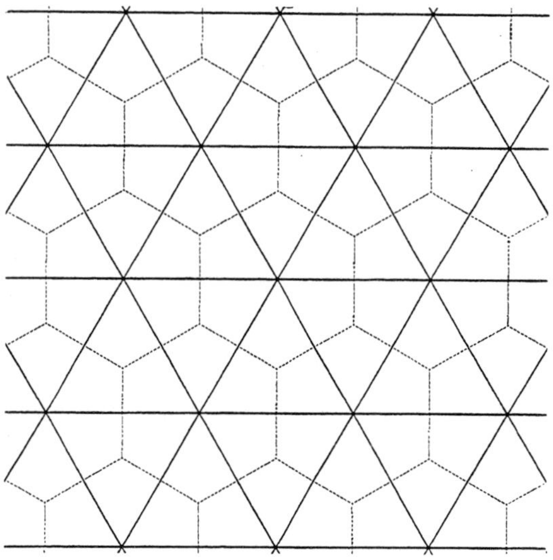

{3,6}, dual of {6,3}

Fig.2: Dual Tilings

TRUNCATED TILINGS

Either one of a dual pair of tilings can be transformed into the other by a process analogous to truncation. This can be achieved by treating each vertex as a growth point for a new face, which as it grows slowly squeezes out the old face until eventually there only remain the new faces centred on the old vertices.

During this process a number of semi-regular tilings are encountered composed of regular polygons only, usually of more than one type.

The first truncation of each regular tiling is shown in Figure 3, along with its isosceles triangle vertex figure and Schläfli notation, t{x,y}.

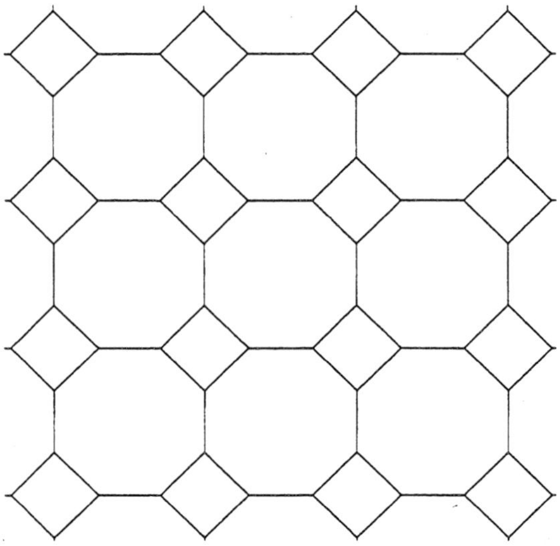

4 2 |4 t{4,4}

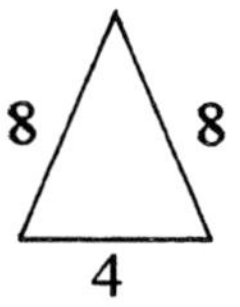

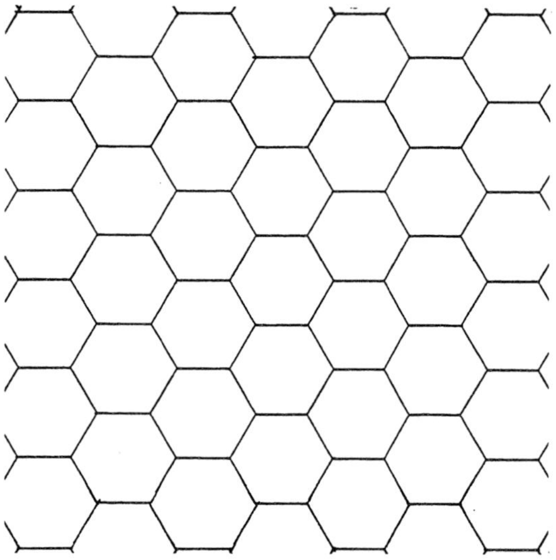

6 2 |3 t{3,6}

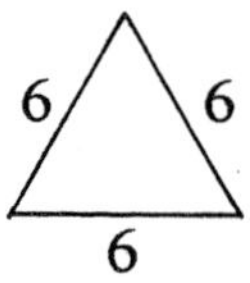

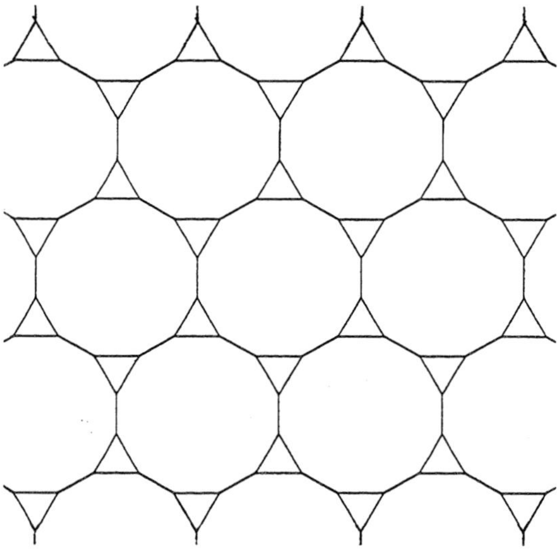

3 2 |6 t{6,3}

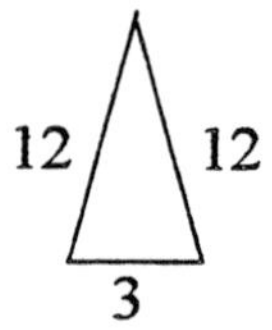

Fig.3: Truncated Tilings

QUASI-REGULAR TILINGS

Continuation of this 'truncation' process leads on to a second truncated tiling, at which point the original polygon faces of the tiling have become their duals, and the new-grown faces meet and touch each other at their vertices.

These are the quasi-regular tilings with two types of face, each surrounded entirely by the other and are given the Schläfli notation $\begin{Bmatrix}x\\y\end{Bmatrix}$.

Because they have two types of face arranged in pairs opposite each other around the vertex, they have rectangular vertex figures. These quasi-regular tilings are shown in Figure 4, where it will be seen that the second truncation of {x,y} is the same as that of {y,x}.

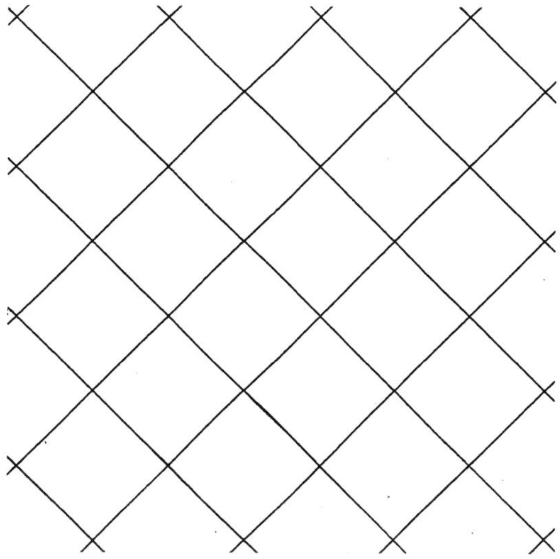

2| 4 4 $\begin{Bmatrix}4\\4\end{Bmatrix}$

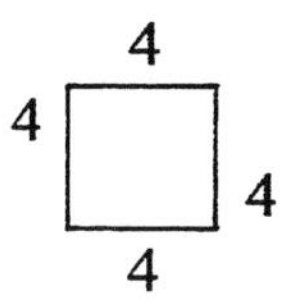

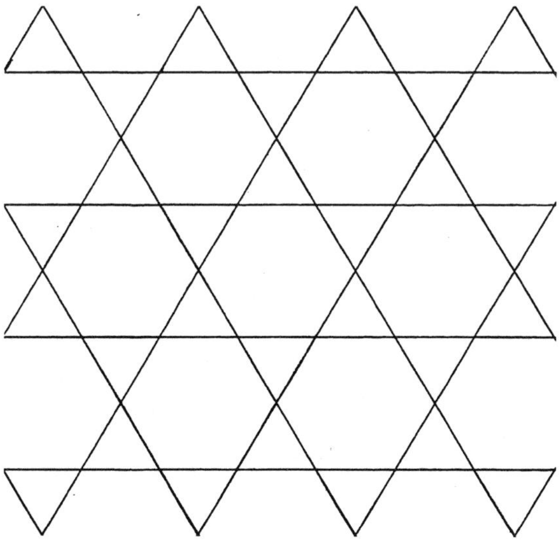

2| 3 6 $\begin{Bmatrix}3\\6\end{Bmatrix}$

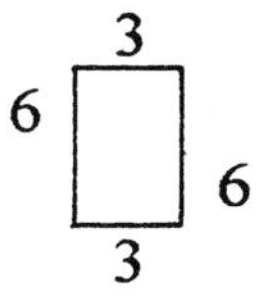

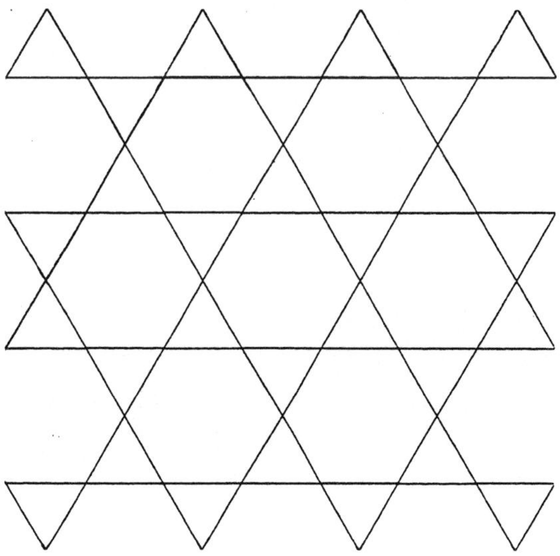

2| 6 3 $\begin{Bmatrix}6\\3\end{Bmatrix}$

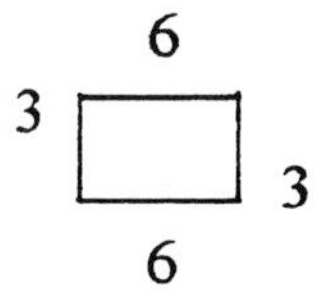

Fig.4: Quasi-Regular Tilings

FIRST TRUNCATION SERIES

The quasi-regular tiling for each dual pair can thus be seen as the centre of a series of truncations that lead from one of the pair to the other.

In fact the first truncation of one can also be regarded as the third truncation of the other, where the new-grown faces have met at their vertices and grown on to form a new edge between themselves, effectively truncating each other. This further squeezes the original faces towards their eventual extinction as points at their centres.

We thus have a series of truncations leading from either of a dual pair (with regular polygon vertex figures) via a truncated semi-regular tiling (with an isosceles triangle vertex figure) to a central quasi-regular tiling (with rectangular vertex figure) and onwards via the other truncation to the other of the pair. Figure 5 shows a typical series of such truncations, the vertex figures of which can all be predicted from the initial pair {x,y}: the only polygons appearing in these being {x} and {y} and their respective truncations {2x} and {2y}. Figure 6 shows this first series for each of our two symmetry families {4,4} and {3,6}.

It will be worth noting for future reference that whilst polygons and their duals form an edge-vertex duality series of three regular forms between one and the other, i.e. (x,2x,x), tilings (and polyhedra) form a face-vertex duality series of five forms consisting of two of these series overlapped at the central quasi-regular form, i.e. (x,2x,x,x,-) and (-,y,y,2y,y). In a similar way we will see in future volumes that closepacks (and hypersolids) form truncation series of semi-regular figures seven forms long as a result of their solid-vertex duality.

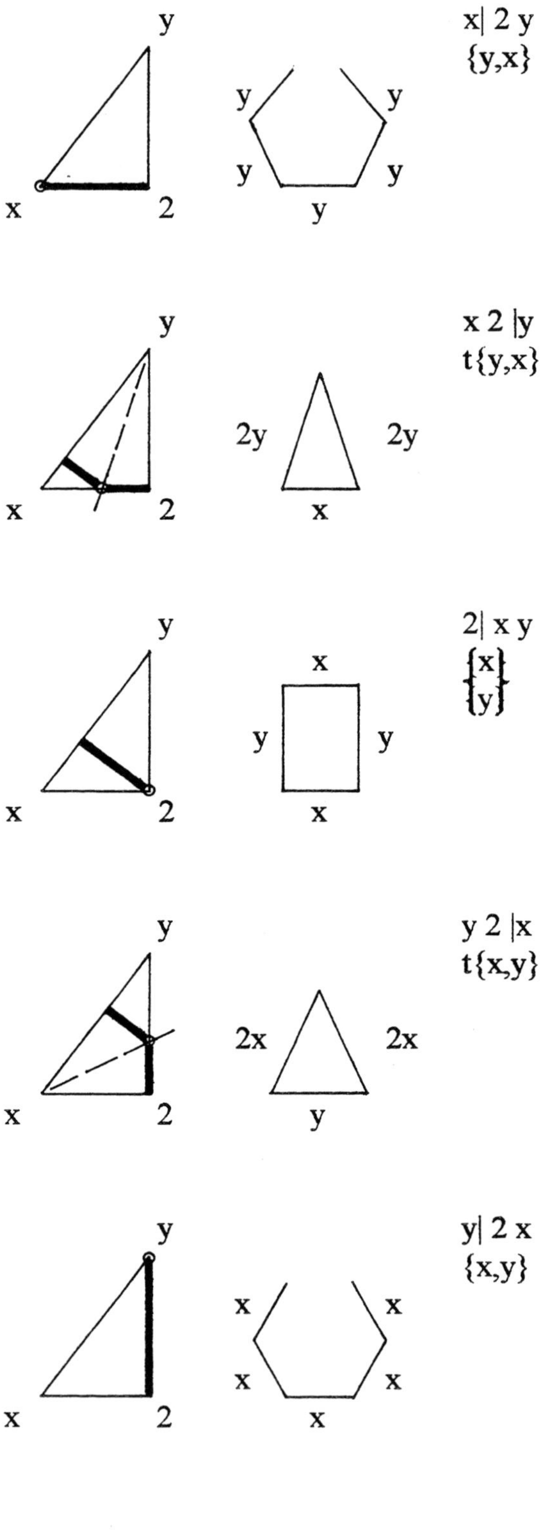

Fig.5: Typical First Series

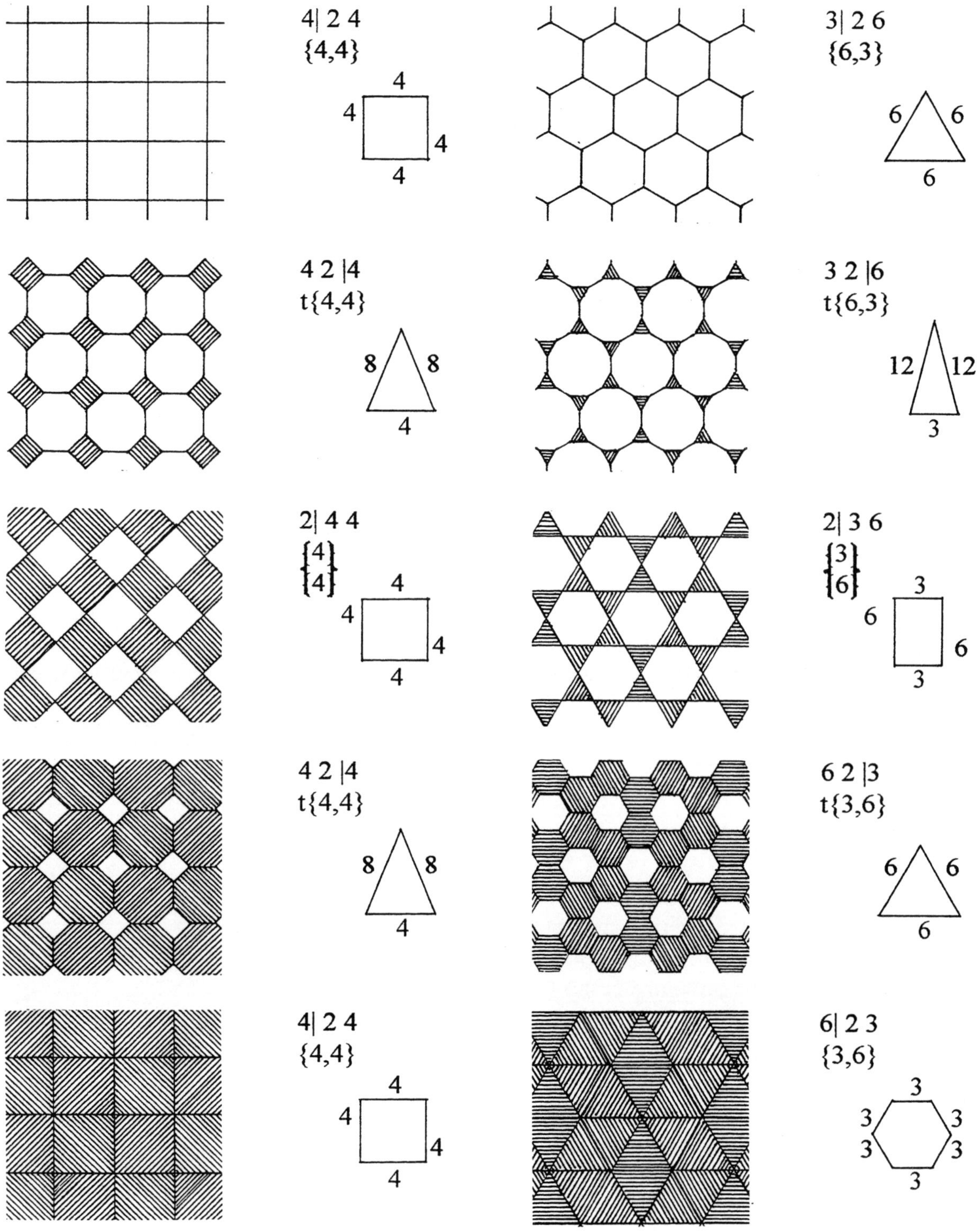

Fig.6: First Series, {4,4} Family

First Series, {3,6} Family

RHOMBIC TILINGS

Taking the quasi-regular tilings and joining together the centres of their faces we can form further dual tilings, this time generally with rhombic faces, as shown in Figure 7.

Taking the centres of the rhombic faces of these tilings will also yield back the original quasi-regular tilings in their turn as can be seen in Figure 8.

A note on vertex figures would seem appropriate at this point. We have thus far been differentiating tilings by reference to the unique arrangement of polygons that occurs at each vertex. This has been summarised in terms of a 'vertex figure', which is basically a polygon joining the mid points of all the edges that meet at any vertex. As we saw in the previous volume each edge of this vertex figure polygon will have a length equal to the 'dual edge length' of the polygon occuring between the two tiling edges chosen. Vertex figures are thus diagrammatic polygons constructed from dual edge lengths.

In the case of rhombic tilings, the faces are not regular polygons and the vertices are generally of two types so that a vertex figure cannot be constructed. However the diagonals of the rhombs form two separate sets of different lengths, proportional to the dual edge lengths of the polygons in the original tilings. Indeed, each set of rhomb diagonals actually delineates one of the parent regular tilings as shown in Figure 9.

In the special case of the self-dual square tiling, both the quasi-regular tiling and its dual the rhombic tiling are square tilings too, so that the rhombs are squares and a vertex figure can be constructed, although it is not shown here.

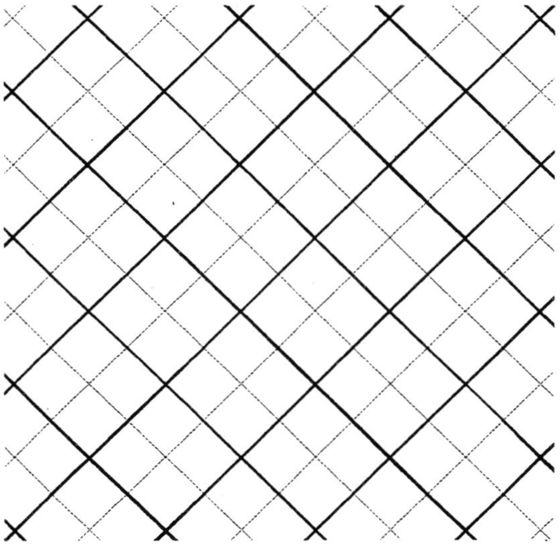

Square Tiling, dual of $\begin{Bmatrix}4\\4\end{Bmatrix}$

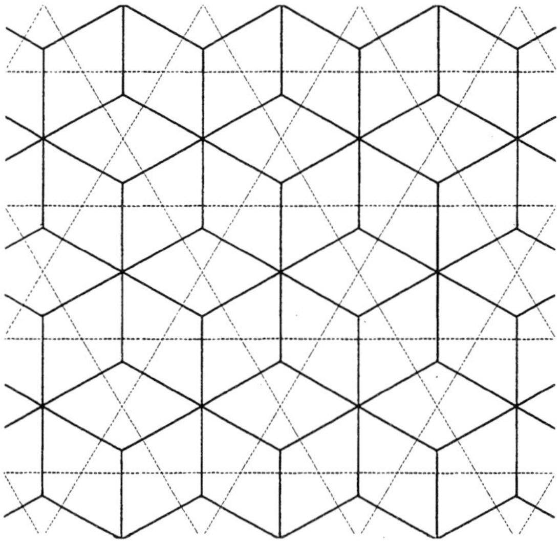

Rhomb Tiling, dual of $\begin{Bmatrix}3\\6\end{Bmatrix}$

Fig. 7: Rhombic Tilings

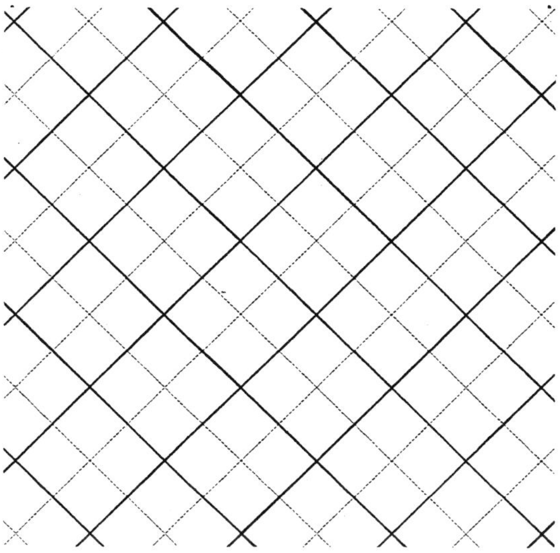

$\begin{Bmatrix}4\\4\end{Bmatrix}$, dual of Square Tiling

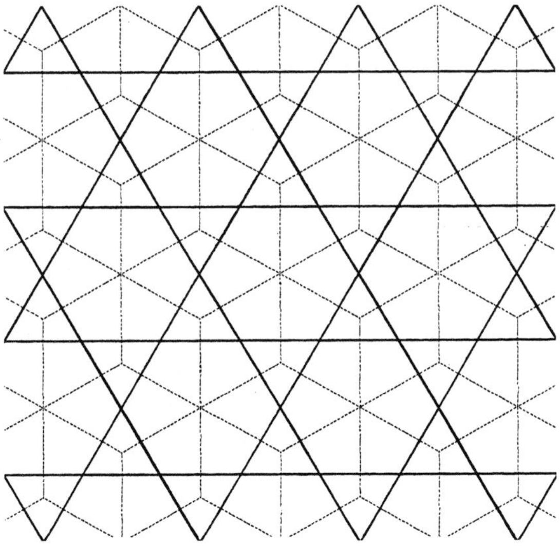

$\begin{Bmatrix}3\\6\end{Bmatrix}$, dual of Rhomb Tiling

Fig.8: Duals of Rhombic Tilings

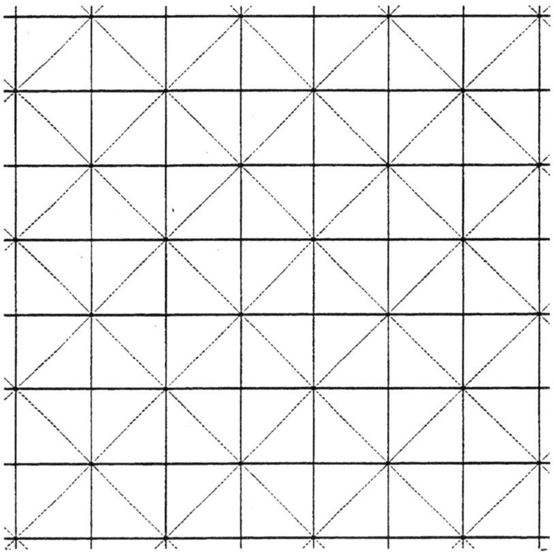

{4,4} and {4,4}

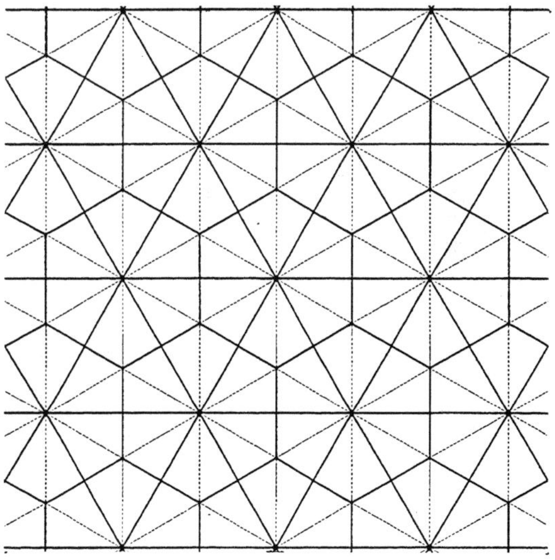

{3,6} and {6,3}

Fig.9: Diagonals of Rhombic Tilings

EVEN FACED TILINGS

If we attempt to 'truncate' a quasi-regular tiling to obtain a rhombic tiling, we can proceed, but will meet no regular forms en route as the new faces grown at each vertex will be generally rectangular. However, by starting with the rhombic tiling and truncating at different rates at the two different types of vertex, we can ensure that the rectangle is normalised into a square resulting in further semi-regular tilings.

The notional third truncation of such a rhombic tiling is therefore equivalent to a slightly distorted first truncation of the quasi-regular tiling. These even faced tilings are so called because generally they have one square face and one truncated face from each parent tiling at each vertex, all having an even number of edges. These are shown in Figure 10 and given the Schläfli notation $t\begin{Bmatrix}x\\y\end{Bmatrix}$.

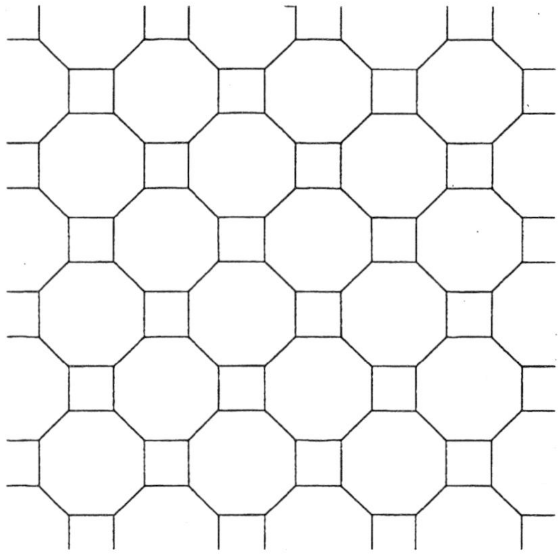

4 2 4| $t\begin{Bmatrix}4\\4\end{Bmatrix}$

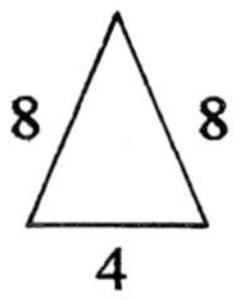

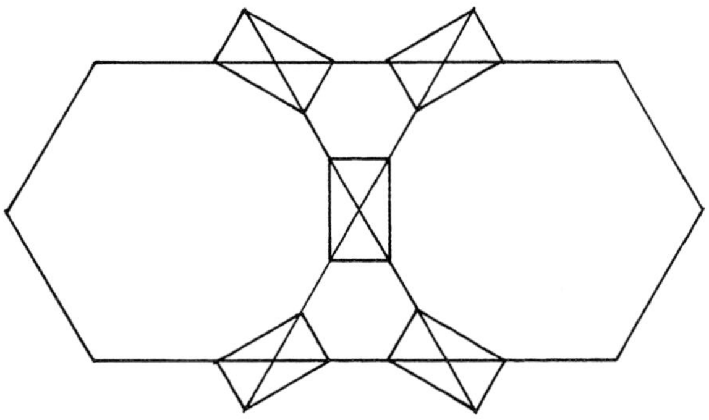

Normalisation required

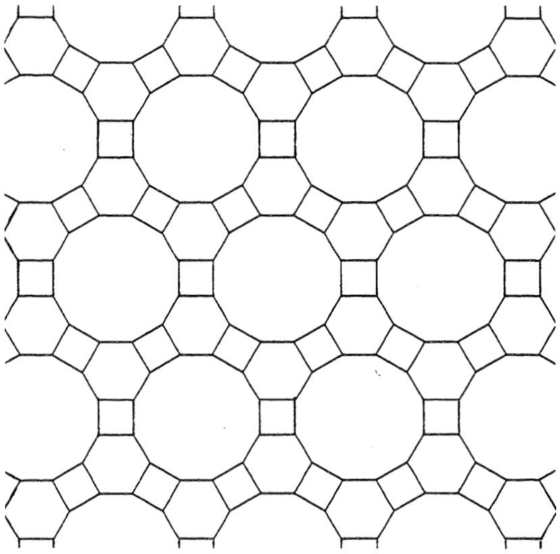

3 2 6| $t\begin{Bmatrix}3\\6\end{Bmatrix}$

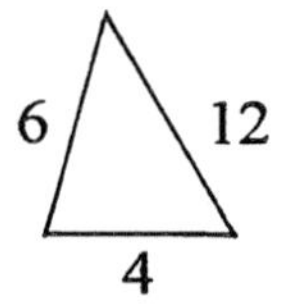

Fig.10: Even Faced Tilings

EVEN MORE TILINGS

In a similar way we can equate the second 'truncation' of a rhombic tiling to a distorted version of the second truncation of a quasi-regular tiling, where again the rectangular faces are normalised into squares.

This second truncation is again semi-regular, somewhat similar to the quasi-regular tiling, but with a four sided bilaterally symmetric vertex figure consisting of the two parent faces opposite each other with two squares in between. These 'rhombi' tilings are shown in Figure 11 and given the Schläfli notation $r\begin{Bmatrix}x\\y\end{Bmatrix}$.

Once more in the case of the {4,4} symmetry family all we get for these first and second 'truncations' are further examples of the square-octagon or square-square tilings.

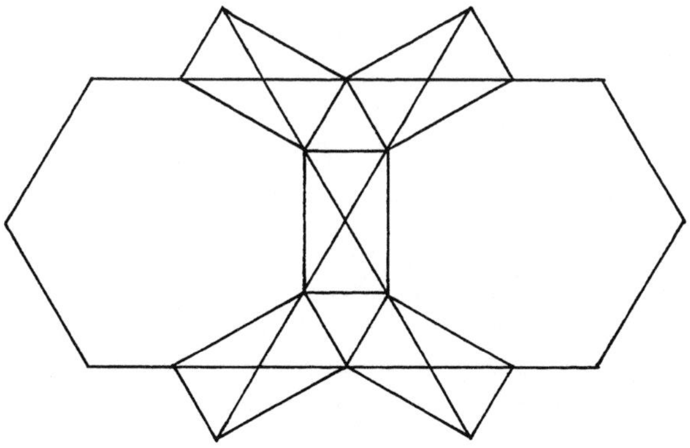

More normalisation

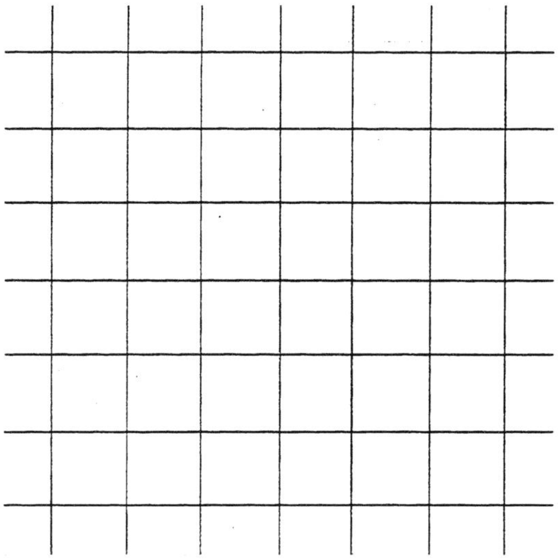

4 4 |2 $r\begin{Bmatrix}4\\4\end{Bmatrix}$

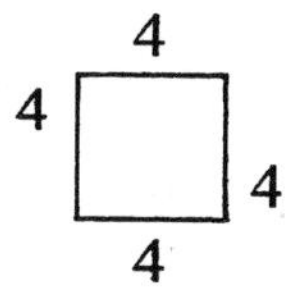

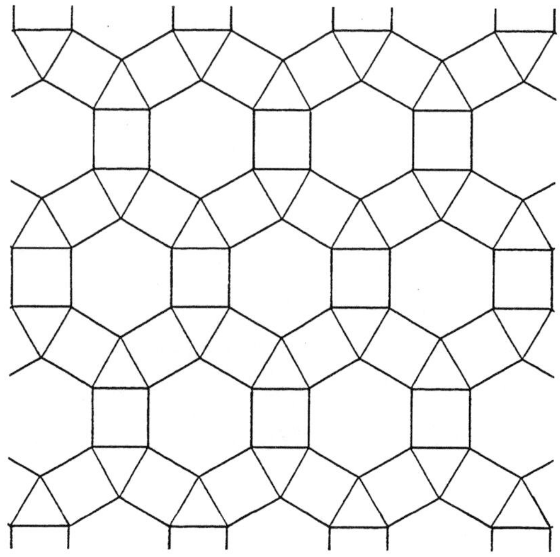

3 6 |2 $r\begin{Bmatrix}3\\6\end{Bmatrix}$

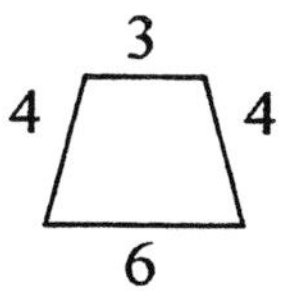

Fig.11: Rhombi Tilings

SNUB TILINGS

One further uniform tiling can be produced for each symmetry family by taking the second truncation of the quasi-regular tiling (the rhombi tiling), leaving out the new square faces and applying a twist until each missing square becomes a 60° rhomb consisting of two equilateral triangles.

The tilings thus produced have a five sided vertex figure with the two parent faces again opposite each other, but now joined by one triangle on one side and two on the other.

The snub tilings are therefore asymmetrical, appearing in apparently left or right handed 'enantiomorphic' forms, when $x \neq y$. This discrepancy can easily be resolved by moving one's point of view from one side of the plane to the other, whereupon the two can easily be seen as one. The snub tilings are shown in Figure 12 and are usually given the Schläfli notation $s\begin{Bmatrix}x\\y\end{Bmatrix}$.

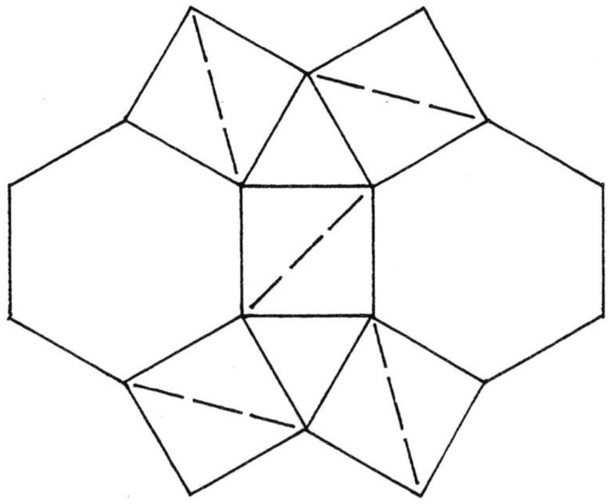

{4} becomes {3} x 2

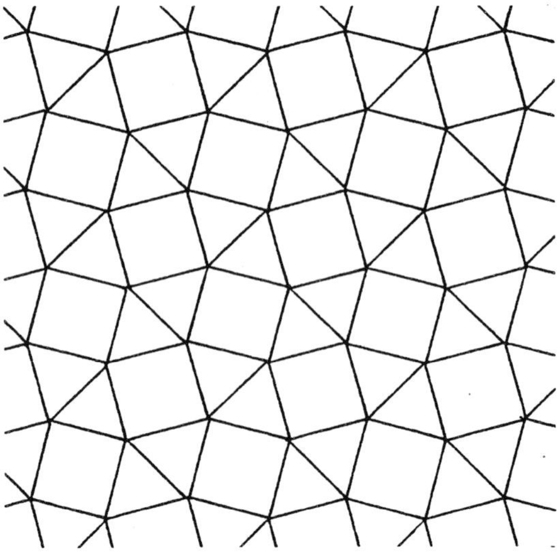

|4 2 4 $s\begin{Bmatrix}4\\4\end{Bmatrix}$

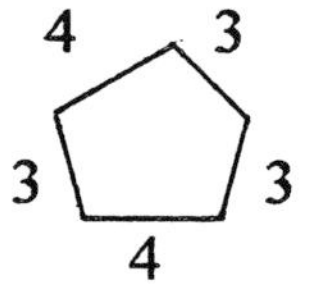

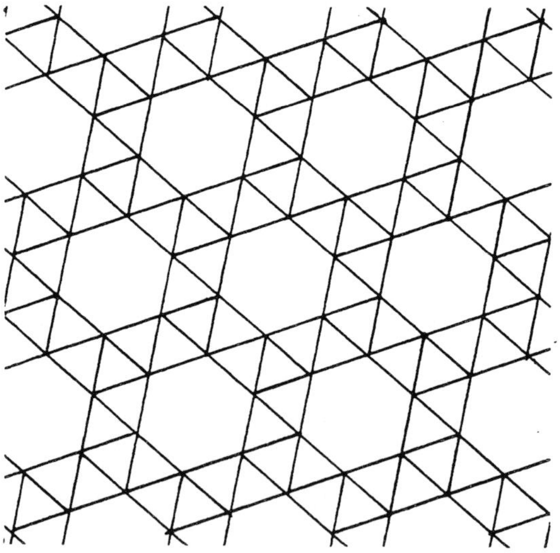

|3 2 6 $s\begin{Bmatrix}3\\6\end{Bmatrix}$

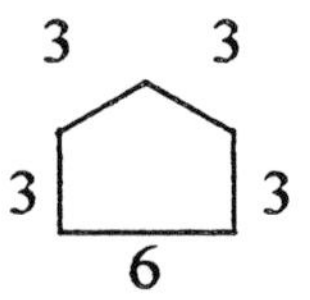

Fig. 12: Snub Tilings

MÖBIUS TRIANGLES

Returning to Figure 9, where the diagonals of the rhombic tilings were seen to delineate the two original tilings of the symmetry family, we have the basis for an alternative and very effective notation for all the tilings encountered thus far. This was used to good effect by Coxeter et al. in their famous paper on 'Uniform Polyhedra' and is equally useful for the current topic of tilings.

Taking the rhombic tiling of either family, we can see that it is entirely composed of small right angled triangles, known as Möbius triangles, as shown in Figure 13. We can label each vertex of these triangles with the symmetry that occurs there, i.e. 6, 4, 3 or 2, and thus by convention label the various points around each triangle that form the vertices of the other members of the same symmetry family. Truncation will thus be seen as a movement of vertices around this triangular grid, semi-regular tilings occurring where vertices are equidistant from the edges of the triangle.

In general we find that specifying the vertices of the Möbius triangle:
y| 2 x = {x,y}, a parent tiling
x| 2 y = {y,x}, the other parent tiling and
2| x y = $\left\{\begin{matrix}x\\y\end{matrix}\right\}$, the quasi-regular tiling;
specifying points on the edges of the Möbius triangle, equidistant from the other edges:
y 2 |x = t{x,y}, a truncated parent tiling
x 2 |y = t{y,x}, a truncated parent tiling and
x y |2 = r$\left\{\begin{matrix}x\\y\end{matrix}\right\}$, the rhombi tiling,
and finally specifying the centre of the Möbius triangle equidistant from all three edges: x 2 y| = t$\left\{\begin{matrix}x\\y\end{matrix}\right\}$, the even faced tiling.
One combination of these symbols remains unused which can conveniently be assigned to the snub tiling, i.e. |x 2 y = s$\left\{\begin{matrix}x\\y\end{matrix}\right\}$.

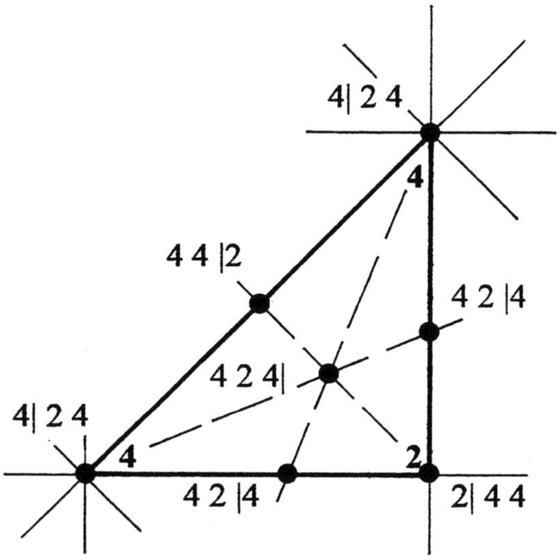

(4 2 4) generating triangle

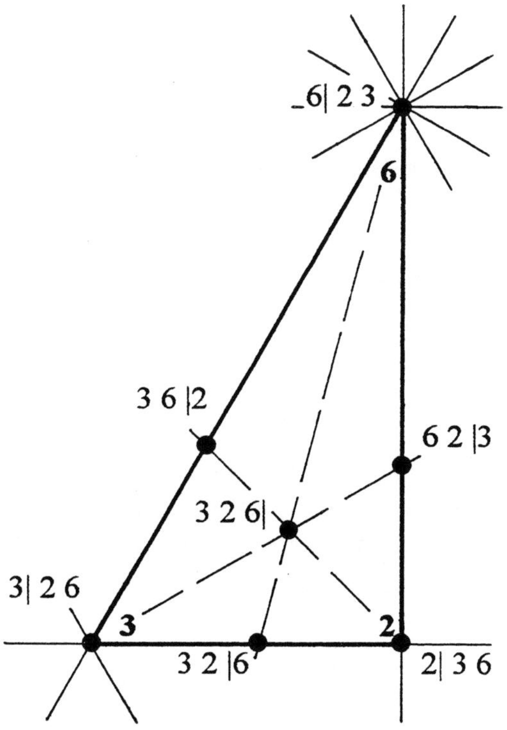

(3 2 6) generating triangle

Fig. 13: Möbius Triangles

A SECOND SERIES

The notional third truncation of a quasi-regular tiling should be a truncated rhombic tiling, however this cannot generally be normalised into a semi-regular form because to get the regular octagon required would entail the parent rhombs themselves being normalised into squares.

This is clearly impossible except in the special case of a self dual family where x = y and the rhomb is a square anyway.

A second series of tilings can however be produced for each dual pair started with, by conveniently inserting the snub tiling in place of the non-existent truncated rhombic tiling for each family.

The series thus comprises a quasi-regular tiling (with a rectangular vertex figure), its 'truncation' the even faced tiling (with a triangular vertex figure), its second 'truncation' the rhombi tiling (with an isosceles trapezium vertex figure), the snub tiling obtained by twisting the vertices of this last, and finally the rhombic tiling itself, dual of the quasi-regular tiling and whose two sets of diagonals delineate the original dual pair.

Figure 14 shows a typical series of such truncations, whilst Figure 15 shows the series for each of our two symmetry families {4,4} and {3,6}.

We have now considered all 10 of the regular and semi-regular tilings which consist of regular polygons filling the plane. These polygons have individually had a density of 1, which value can be assigned to the entire tiling as a single layer of such polygons.

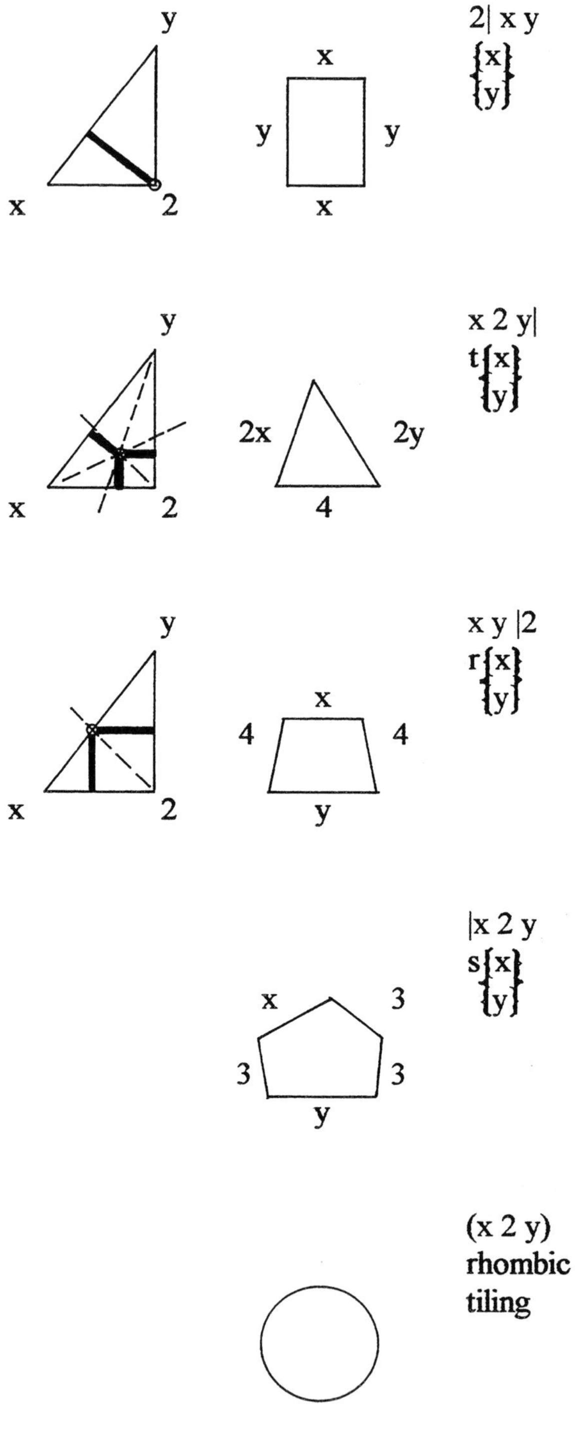

Fig.14: Typical Second Series

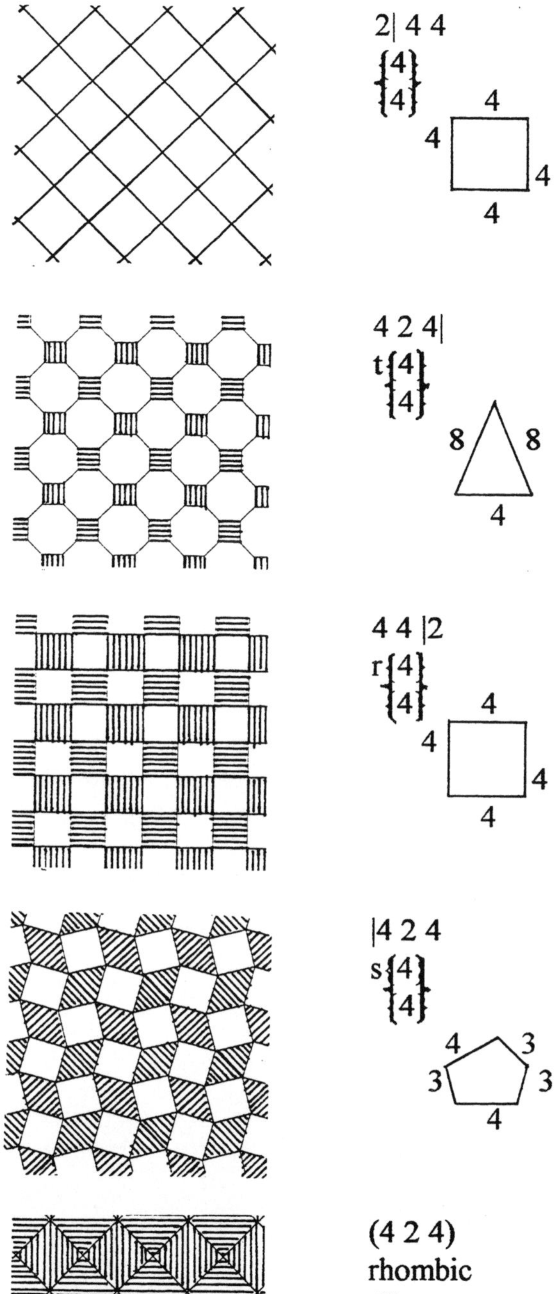

Fig.15: Second Series, {4,4} Family

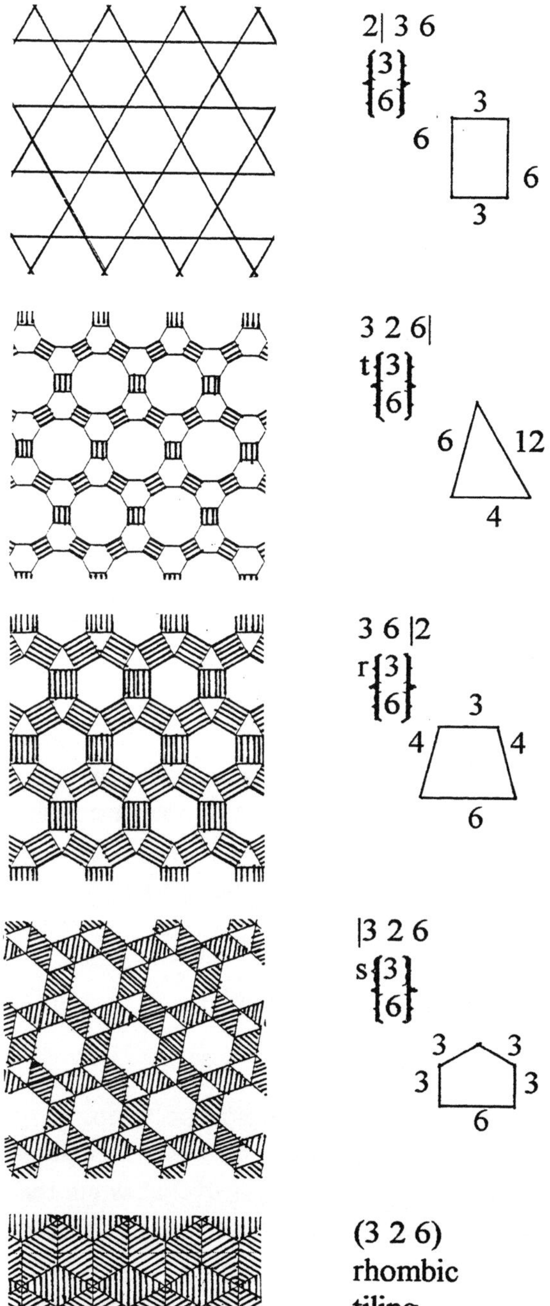

Second Series, {3,6} Family

QUASITRUNCATION GENERALLY

Thus far we have been 'truncating' tilings by growing new faces at each vertex. This has caused the original polygon faces of each tiling to become their truncations and then their duals in turn, producing semi-regular tilings along the way. These have involved the polygons {6}, {4} and {3} along with their truncations {12}, {8} and {6}.

A polygon can also be quasitruncated by treating it as its inverse and then truncating, this process leading to both a quasitruncated polygon and then the original polygon's dual once again. The quasitruncations of the polygons {6}, {4} and {3} are truncations of {6/5}, {4/3} and {3/2} respectively, i.e. the conjugate polygons {12/5}, {8/3} and {²3/1} as shown in Figure 16. Two of these are stars and the last a double polygon, with densities of 5, 3 and 2 respectively.

Tilings can thus be quasitruncated by growing a new face based on the original vertex figure in the normal way, but then substituting the quasitruncated polygon for the truncated one. We can thus assemble a number of new tilings with appropriate vertex figures consisting of new grown faces and quasitruncated faces.

It should be noted that star polygons have densities greater than 1 and their edges seem to cut across each other, so that when they share an edge their faces overlap with their neighbours to some extent. Tilings with star polygons thus have apparently increased density in the plane and can be considered as multi-layered tilings, not strictly tilings in the mathematical sense, nor realisable in the everyday sense as tilings on a wall, but nevertheless achievable in the zero thickness of our two dimensional domain.

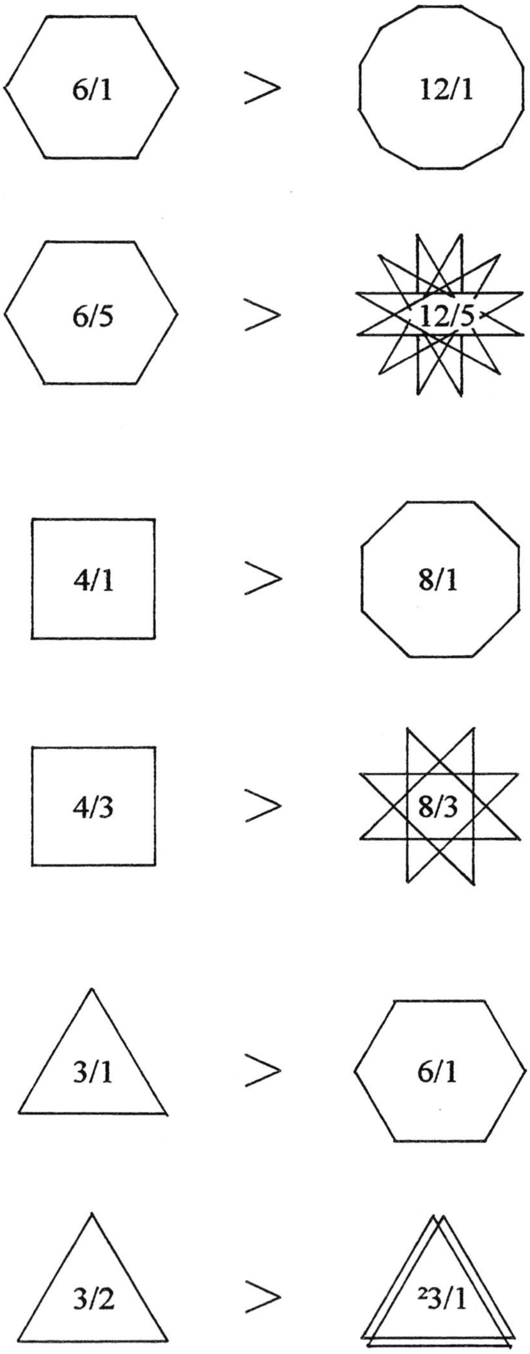

Fig.16: Quasitruncating Polygons

QUASITRUNCATED TILINGS

Applying the quasitruncation process to our original three regular tilings we obtain three quasitruncated regular tilings. These are shown in Figure 17, along with their vertex figures and Schläfli notation, t{x',y}.

The pair of these involving star polygons (quasitruncated {6,3} and {4,4}) can be seen to have exactly the same grid arrangement of vertices as our original truncated tilings, however the vertices are now further away from the centres of the faces on account of the stellation involved.

The quasitruncated {3,6} differs on account of the double polygon involved having doubled up vertices and edges which cannot be produced by stellating the truncation. It does however share its vertices with the parent figure being truncated.

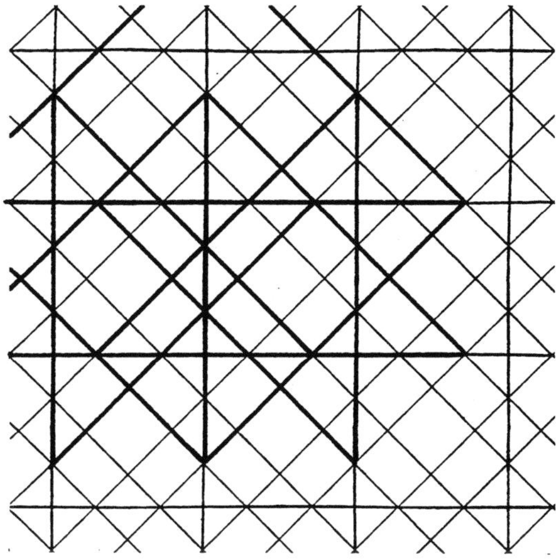

4 2 |4/3   t{4',4}

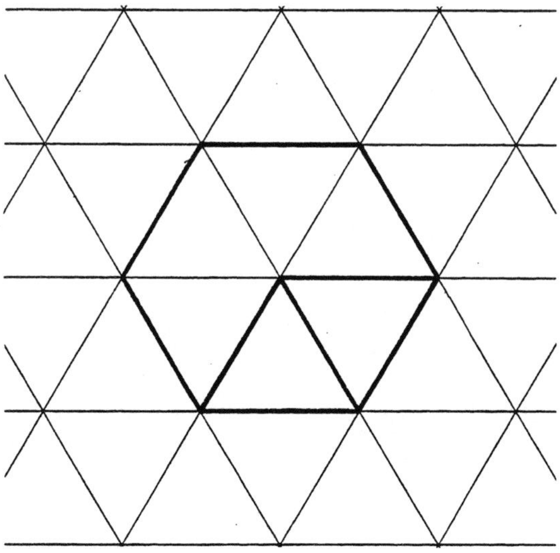

6 2 |3/2   t{3',6}

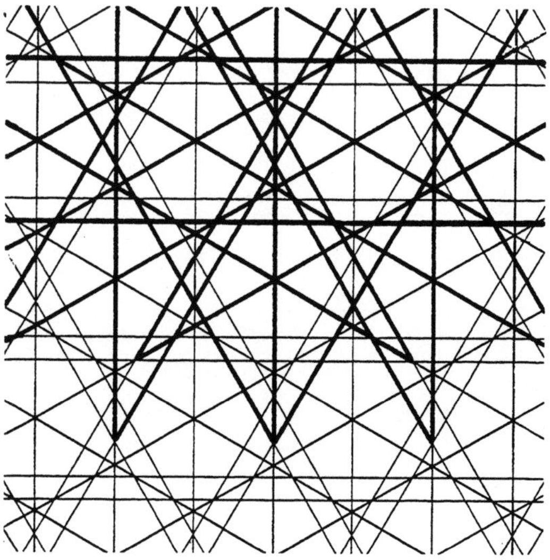

3 2 |6/5   t{6',3}

Fig. 17: Quasitruncated Tilings

QUASI EVEN FACED TILINGS

Continuation of the quasitruncation process to a single polygon leads on to the dual of that polygon, i.e. itself again. Thus the second 'quasitruncation' of a regular tiling will in fact be identical to the quasi-regular tiling already described. It will however have two types of face, one pair of regular faces and one pair of inverse faces, each pair opposite each other on the vertex figure.

This quasi-regular tiling can be 'truncated' to produce a further quasitruncation for each symmetry family, equivalent to the even faced tiling, but with the quasitruncated polygons in lieu of the truncated, and again all normalised to provide squares as the new faces rather than rectangles.

These quasitruncated quasi-regular tilings are shown in Figure 18 and are given the Schläfli notation $t\begin{Bmatrix}x'\\y\end{Bmatrix}$.

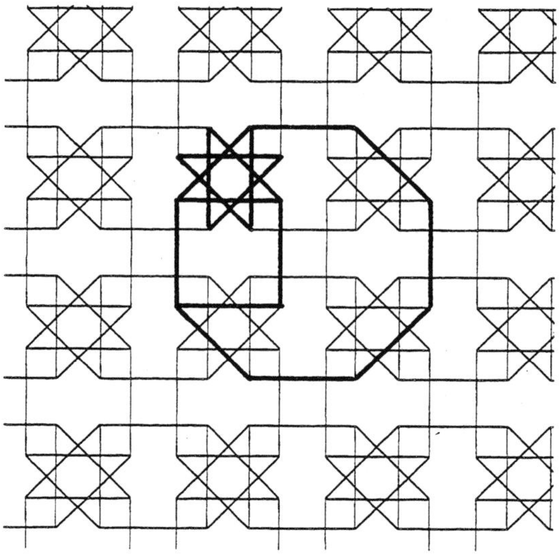

4 2 4/3| $t\begin{Bmatrix}4'\\4\end{Bmatrix}$

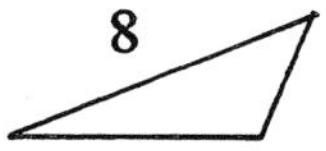

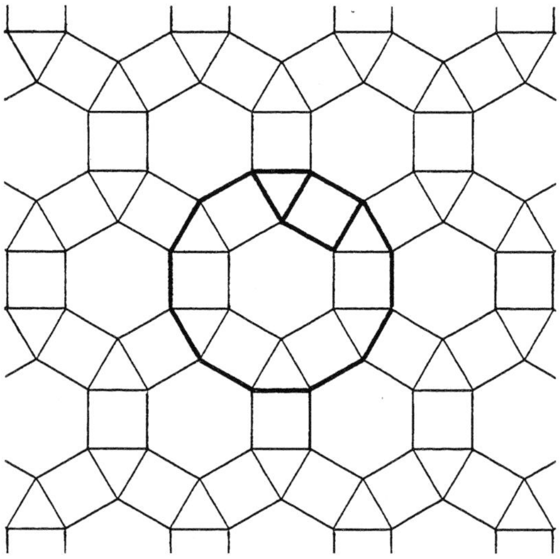

3/2 2 6| $t\begin{Bmatrix}3'\\6\end{Bmatrix}$

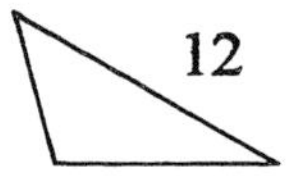

3 2 6/5| $t\begin{Bmatrix}6'\\3\end{Bmatrix}$

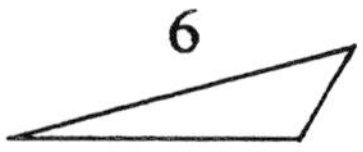

Fig. 18: Quasi Even Faced Tilings

QUASIQUASITRUNCATED TILINGS

As with the quasitruncated regular tilings, the pair of quasitruncated quasi-regular tilings involving star polygons can be seen to have exactly the same grid arrangement of vertices as our original even faced tilings.

The quasitruncation involving the double polygon is however different, but again shares vertices with another member of the family. In fact it appears to be the same as the rhombi tiling, but has holes where the hexagons fit, albeit covered by dodecagons.

Although we cannot obtain it by truncation or quasitruncation, it is possible to conceive a quasi-regular vertex figure consisting of both pairs of opposing polygons as inverses. Truncating such a figure will produce a kind of double quasitruncation, which is still even faced and will have a predictable vertex figure. This can be constructed and is given the Schläfli notation $t\begin{Bmatrix}x'\\y'\end{Bmatrix}$.

This quasiquasitruncated even faced tiling for each symmetry family is shown in Figure 19.

It may have been noticed with the foregoing quasitruncated tilings that the vertex figures predicted do not seem to fill the 360° required at a vertex. Generally however the internal angle of the largest of the three polygons is filled by the internal angles of the two smaller ones.

Interestingly, with all these quasitruncations, if the larger polygon is assigned positive density, and the smaller ones negative, then the resultant density taken anywhere over the entire tiling (and adjusted for the increased density of star or double polygons as appropriate) remains at 1.

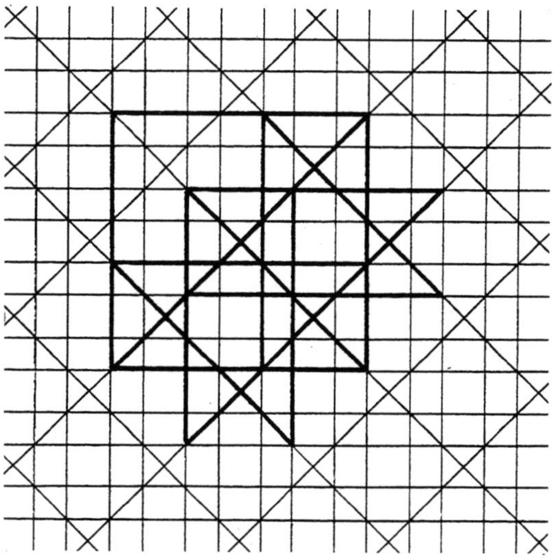

4/3 2 4/3| $t\begin{Bmatrix}4'\\4'\end{Bmatrix}$

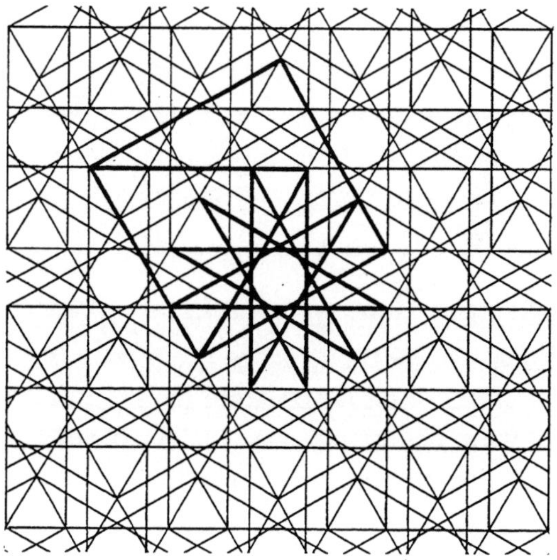

3/2 2 6/5| $t\begin{Bmatrix}3'\\6'\end{Bmatrix}$

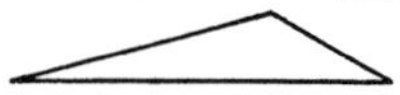

Fig.19: Quasiquasitruncated Tilings

QUASIRHOMBI TILINGS

Starting with the quasi-regular tilings and continuing the quasitruncation process beyond the quasi even faced tilings leads to a second truncation with two square faces and two original faces at each vertex.

We have already seen the rhombi tilings resulting from normal truncation, but in the case of quasitruncation one of the faces in the predicted trapezium vertex figure will be an inverse polygon. This polygon can be turned inside out, or inverted back into a regular polygon, by twisting the vertex figure so that the two square faces now intersect on the diagonals.

These quasirhombi tilings are shown in Figure 20 and are given the Schläfli notation $r'\begin{Bmatrix}x\\y\end{Bmatrix}$.

The quasirhombi tiling in the {4,4} family is interesting in that the quasitruncation process so folds up the vertices that all the squares comprising the tiling are coincident. It therefore contains an infinite number of squares (sufficient to fill the plane) but all in one place and perhaps should not be regarded as a tiling at all.

The quasirhombi tiling in the {3,6} family appears to be the same as the quasiquasi-truncated tiling in the same family, but differs in not having any {12/5} faces, just holes left where they would fit.

Performing the same procedure of continuing the quasitruncation with the quasiquasitruncated tiling, we end up back with the original rhombi tiling. With two inverse polygons opposite each other in the vertex figure, it needs twisting twice to remove the inverses, which leaves us exactly where we started.

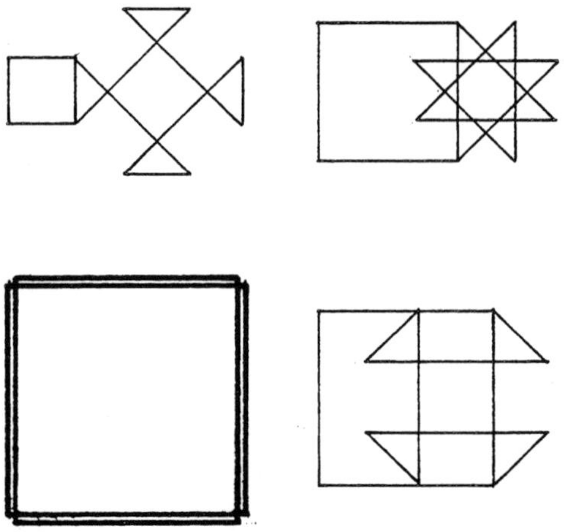

4/3 4 |2 $r'\begin{Bmatrix}4\\4\end{Bmatrix}$

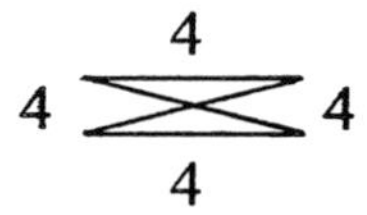

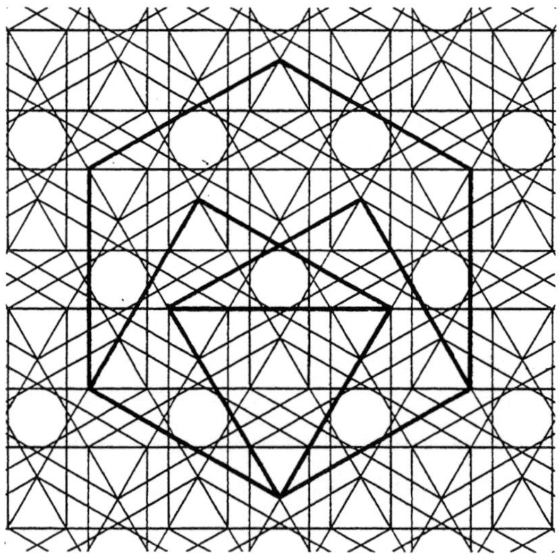

3/2 6 |2 $r'\begin{Bmatrix}3\\6\end{Bmatrix}$

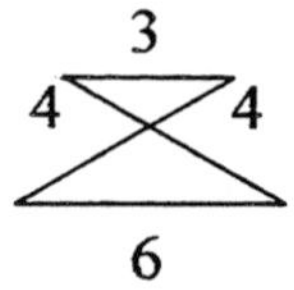

Fig.20: Quasirhombi Tilings

QUASISNUB TILINGS

Whilst the rhombi tilings have two missing faces on their vertex figure diagonals, the new quasirhombi tilings with two squares on the diagonals have two missing faces on the perimeter of the vertex figure. The missing faces in these two tilings turn out to be conjugate polygons: {12} and {12/5} in the {3,6} family and {∞} and {2} in the {4,4}.

Taking these quasirhombi tilings in the same way we took the rhombi tilings, we can once again leave out the new square faces and apply a twist, replacing each square with two equilateral triangles.

The tilings produced once again have five sided vertex figures, but this time more twisted up on account of where we started from. They also occur as enantiomorphic pairs with theoretically left and right handed twists when $x \neq y$.

These quasisnub tilings are shown in Figure 21 and are usually given the Schläfli notation $s'\begin{Bmatrix}x\\y\end{Bmatrix}$.

Fortunately the quasisnub tiling in the {4,4} family does not suffer from the restrictions we have just seen with the quasirhombi tiling. With the extra twist it does in fact unfold again across the plane, interestingly sharing the same grid of vertices as the original snub tiling in that family. Here, however, each edge is nearly four (2+√3) times longer and subdivides the original triangles into six, and squares into four.

Comparing the quasisnub tiling in the {3,6} family in the same way with its snub, a similar effect is seen filling in the hexagons, creating a new central vertex, however here the edge length remains the same.

|4 2 4/3 $s'\begin{Bmatrix}4\\4\end{Bmatrix}$

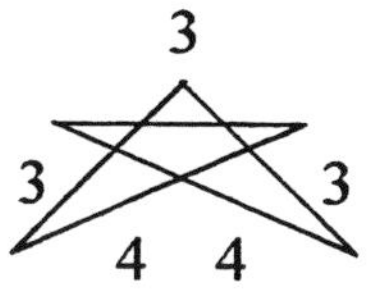

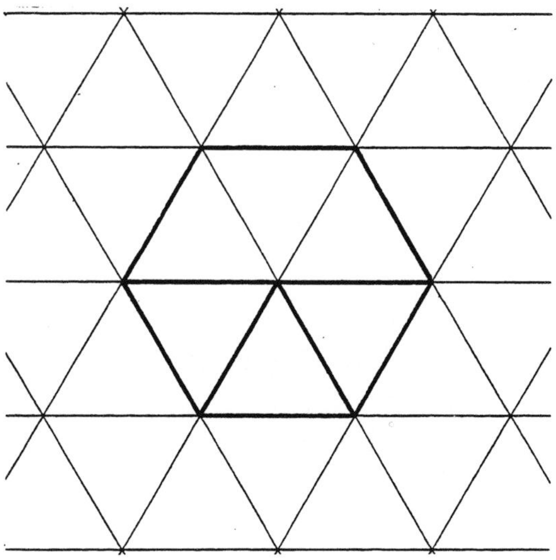

|3/2 2 6 $s'\begin{Bmatrix}3\\6\end{Bmatrix}$

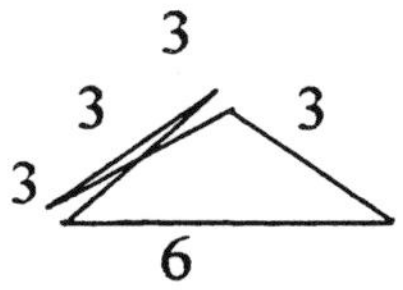

Fig.21: Quasisnub Tilings

DOUBLE POLYGON VERTICES

The observed correspondence between some of the 'quasi' even faced tilings and the related rhombi and quasirhombi tilings in the {3,6} family can be explained in terms of double polygons. Within these particular even faced tilings, double polygons occur in the vertex figures, generally a {^{2}3/1}, resulting from the truncation of a {3/2}.

When there is such a double polygon in a triangular vertex figure, it can be regarded as a single polygon occurring twice there. This leads to a double triangular vertex figure with six edges, where the vertex is effectively visited twice to complete it, as shown in Figure 22.

The result shown is a quadrilateral shaped vertex figure with one face missing, but its diagonals included. This will thus produce the same pattern in the plane as the quadrilateral vertex figure of the associated rhombi or quasirhombi tiling. We will see much later that we can only obtain a quadrilateral vertex figure in this way when a double polygon is involved (see p.72).

In the particular case where two double polygons occur in a triangular vertex figure (t{3',6}), the above process only partly solves the problem, leaving new double faces unresolved. Treating these and their further products in turn in the same manner will fortunately resolve the matter, changing all double polygons to twice visited single ones and completing a full circle with the vertex figure as shown. Extracting the central figure from the vertex we have just produced, we have a vertex figure for a tiling that appears to be a {6,6/2}.
If this has a dual in the form {6/2,6}, we may have a new family of tilings to consider: a hint of things to come!

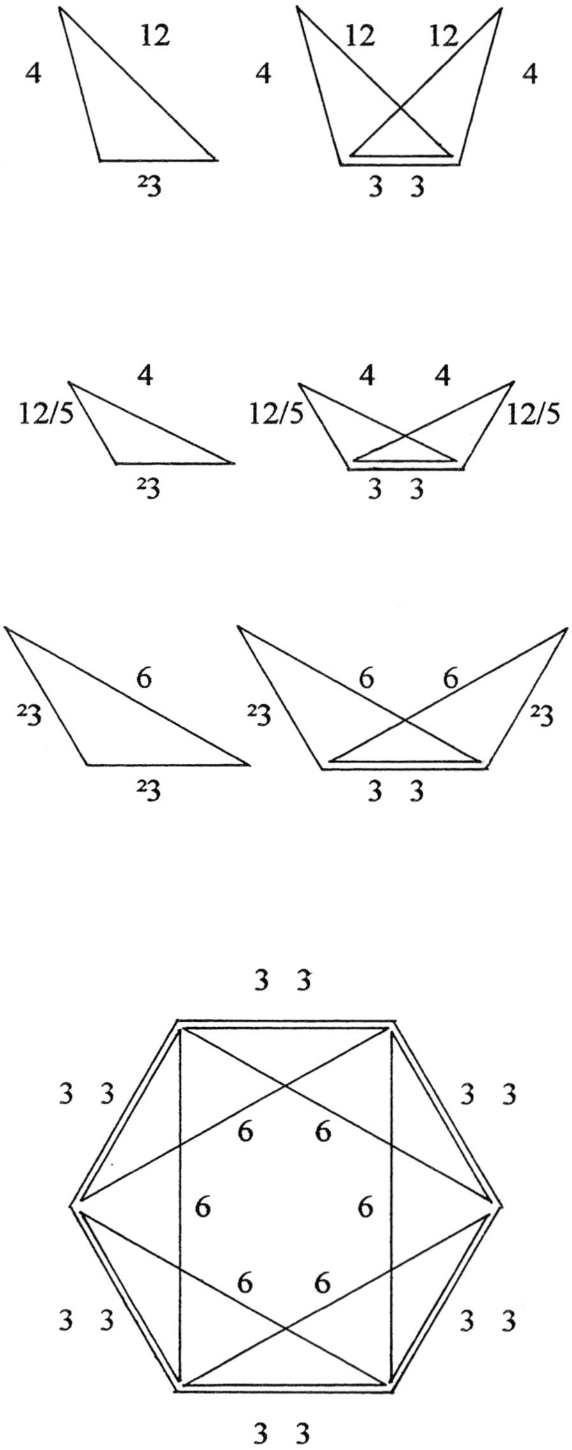

Fig.22: Double Polygon Vertex Figures

VERTEX FIGURES AND CYCLES

The quadrilateral vertex figures of the rhombi and quasirhombi tilings can also be drawn in other ways which have not yet been examined. These take the overall six faced vertex figure, a quadrilateral complete with two diagonals, and leave out pairs of faces to get two further descriptions of the same tilings as shown in Figure 23.

Missing out the square faces we can define the p-type rhombi and quasirhombi tilings which can be given the Schläfli notations $p\begin{Bmatrix}x\\y\end{Bmatrix}$ and $p'\begin{Bmatrix}x\\y\end{Bmatrix}$.
Missing out the parent faces we can define the q-type rhombi and quasirhombi tilings which can be given the Schläfli notations $q\begin{Bmatrix}x\\y\end{Bmatrix}$ and $q'\begin{Bmatrix}x\\y\end{Bmatrix}$.
It should be noted that these procedures do not work for the rectangular vertex figures of the quasi-regular tilings (or indeed the {4,4} rhombi tiling) as the diagonals of these are effectively infinite polygons.

Whilst these new tilings are technically different from each other they still share the same patterns in the plane, a property that will not hold when we come to perform the same operation on polyhedral vertex figures.

Figure 23 also shows how the trapezoid rhombi and quasirhombi vertex figures can each be broken down into two triangles. These turn out to be the triangular vertex figures of the four {3,6} family even faced tilings (one normal, two quasitruncated and one quasiquasitruncated). These tilings, by continuing the process of truncation or quasitruncation, lead back again to the rhombi or quasirhombi forms, thus closing the cycle within the family between these various vertex figure configurations.

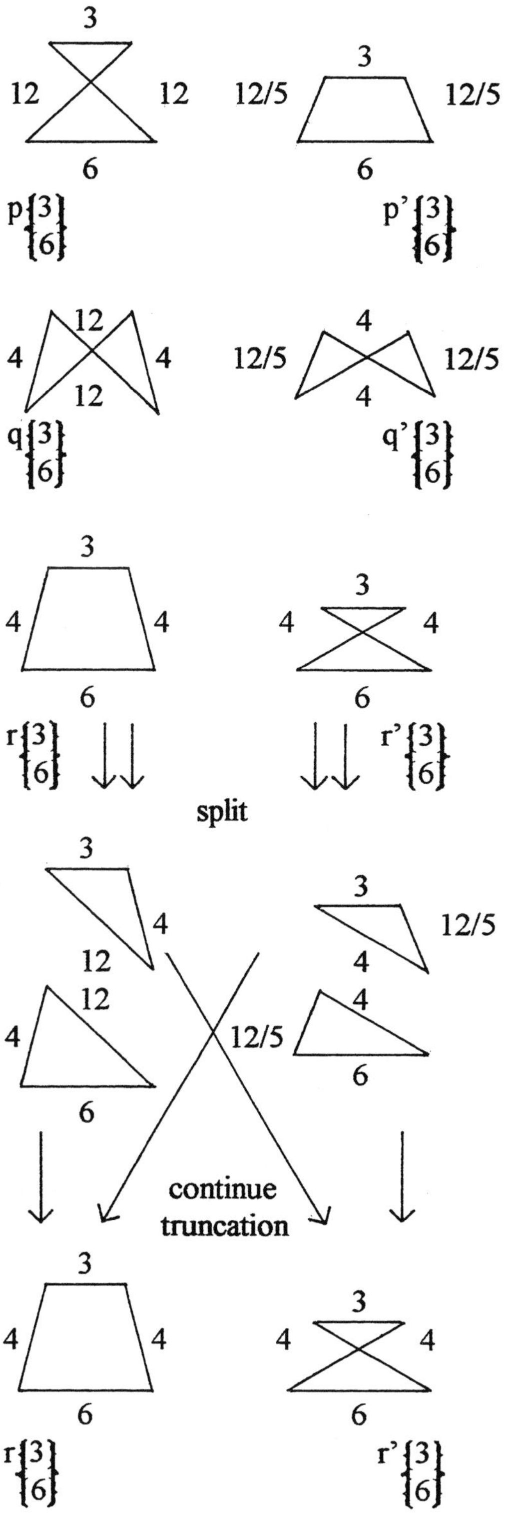

Fig.23: Rhombi Vertex Figures and Cycles

MANIPULATING VERTICES

We have already seen that each of the even faced tilings can be regarded as the centre of a second truncation series which takes {4} as the truncating face of the rectangular vertex of a quasi-regular tiling leading to the four faced trapezoid vertices of the rhombi and quasirhombi tilings.

Considering the other 'even' faces as truncating polygons in their own right we arrive at a number of new short truncation series for our two symmetry families as shown in Figure 24. Interestingly at each end of the series we find a null {2} face, needed to generate its truncation the {4}. The results of quasitruncating the same starting figures are also shown, but none of them generate anything new. What we have here are two new routes across the Möbius triangle, truncating an original parent tiling to obtain the truncation of its dual!

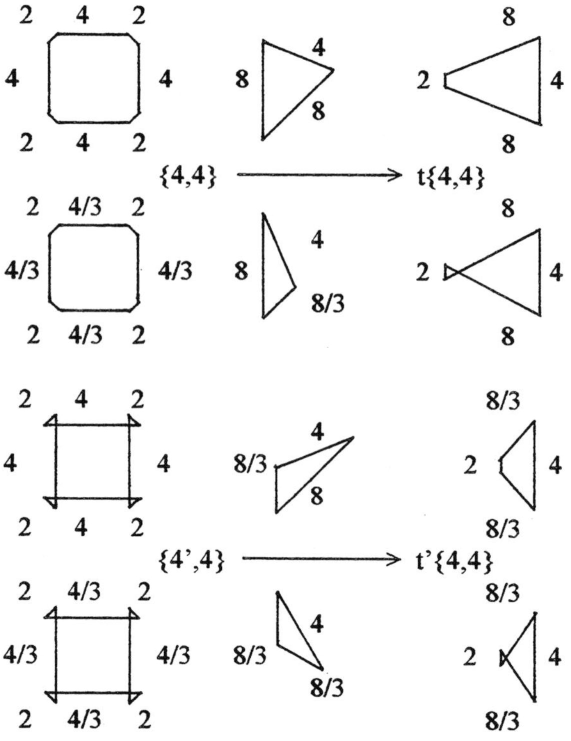

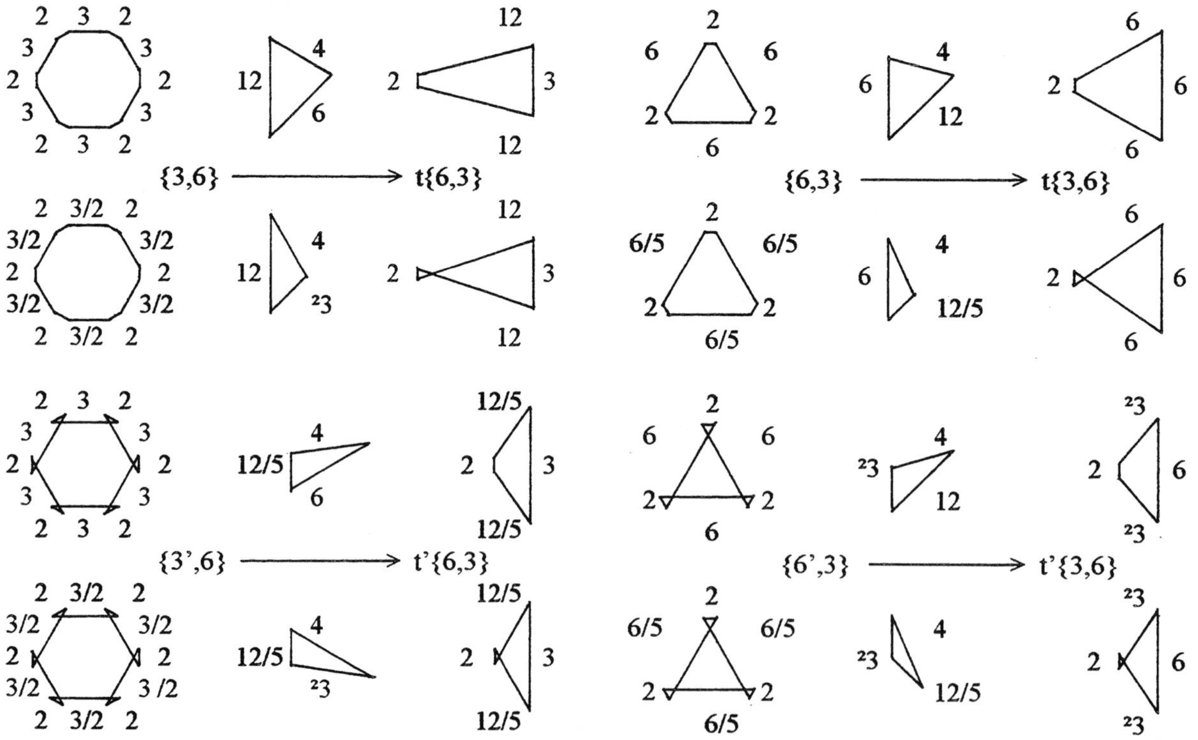

Fig.24: Short Truncation Series

MÖBIUS REVISITED

Thus within each of the original Möbius triangles it is possible to find short series of truncations proceeding from any chosen vertex of the Möbius triangle (a form of quasi-regular tiling) towards the edge opposite (a form of rhombi tiling). In doing this we normalise 'quasi-regular' vertex figures with even numbers of edges into their equivalent regular polygons, even if the faces included are only digons {2}.

We have in the last few pages considered several 'quasi' tilings for which alternative notations have been given in Coxeter's form. This is done by specifying theoretical new generating triangles, which will be known as colunar triangles in the spherical geometry of the polyhedra. These are shown in Figure 25 lying outside the original Möbius triangle. Vertex positions for these 'quasi' forms are again found at points equidistant from the edges of the triangle, but now the edges have to be extended to accommodate this.

In general the tilings we have examined are: specifying points on the edges of the external triangles (x' 2 y), (x 2 y') or (x' 2 y'), equidistant from the other edges:

y 2 |x' = t{x',y}, a quasitruncated tiling

x 2 |y' = t{y',x}, a quasitruncated tiling and

x' y |2 = $r'\begin{Bmatrix}x\\y\end{Bmatrix}$, the quasirhombi tiling

(N.B. x' y' |2 = x y |2, the rhombi tiling); specifying the centres of these new triangles, equidistant from all three edges:

x' 2 y| = $t\begin{Bmatrix}x'\\y\end{Bmatrix}$, a quasi even faced tiling

x 2 y'| = $t\begin{Bmatrix}y'\\x\end{Bmatrix}$, a quasi even faced tiling

x' 2 y'| = $t\begin{Bmatrix}x'\\y'\end{Bmatrix}$, the quasiquasitruncated even faced tiling; and finally we have

|x' 2 y = $s'\begin{Bmatrix}x\\y\end{Bmatrix}$, the quasisnub tiling.

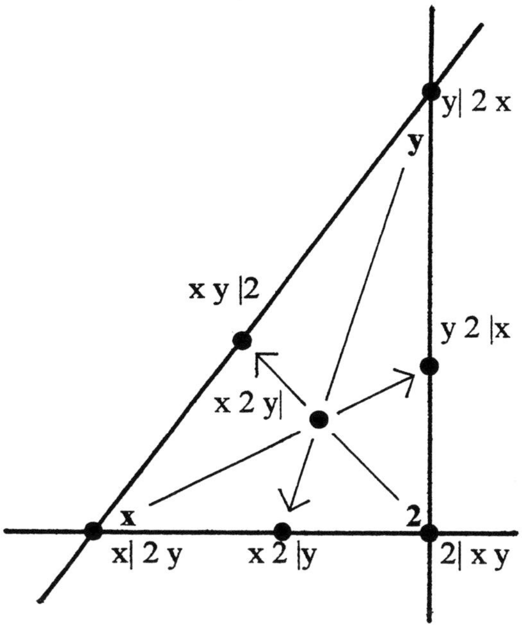

Short Truncation Series

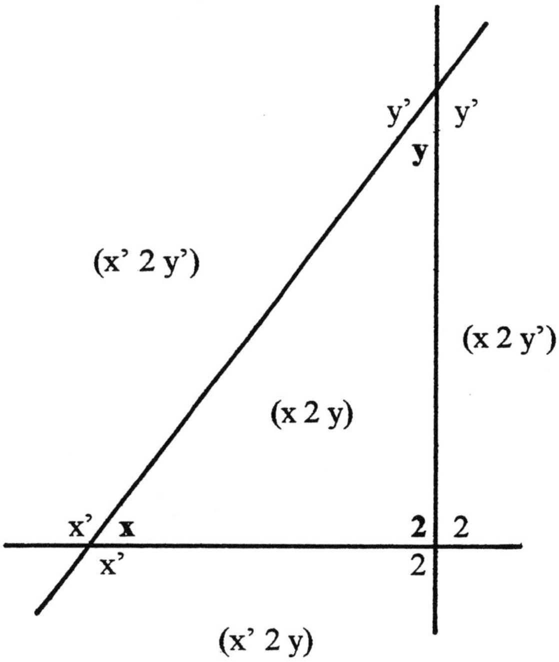

'Colunar' Triangles

Fig.25: Möbius Triangles

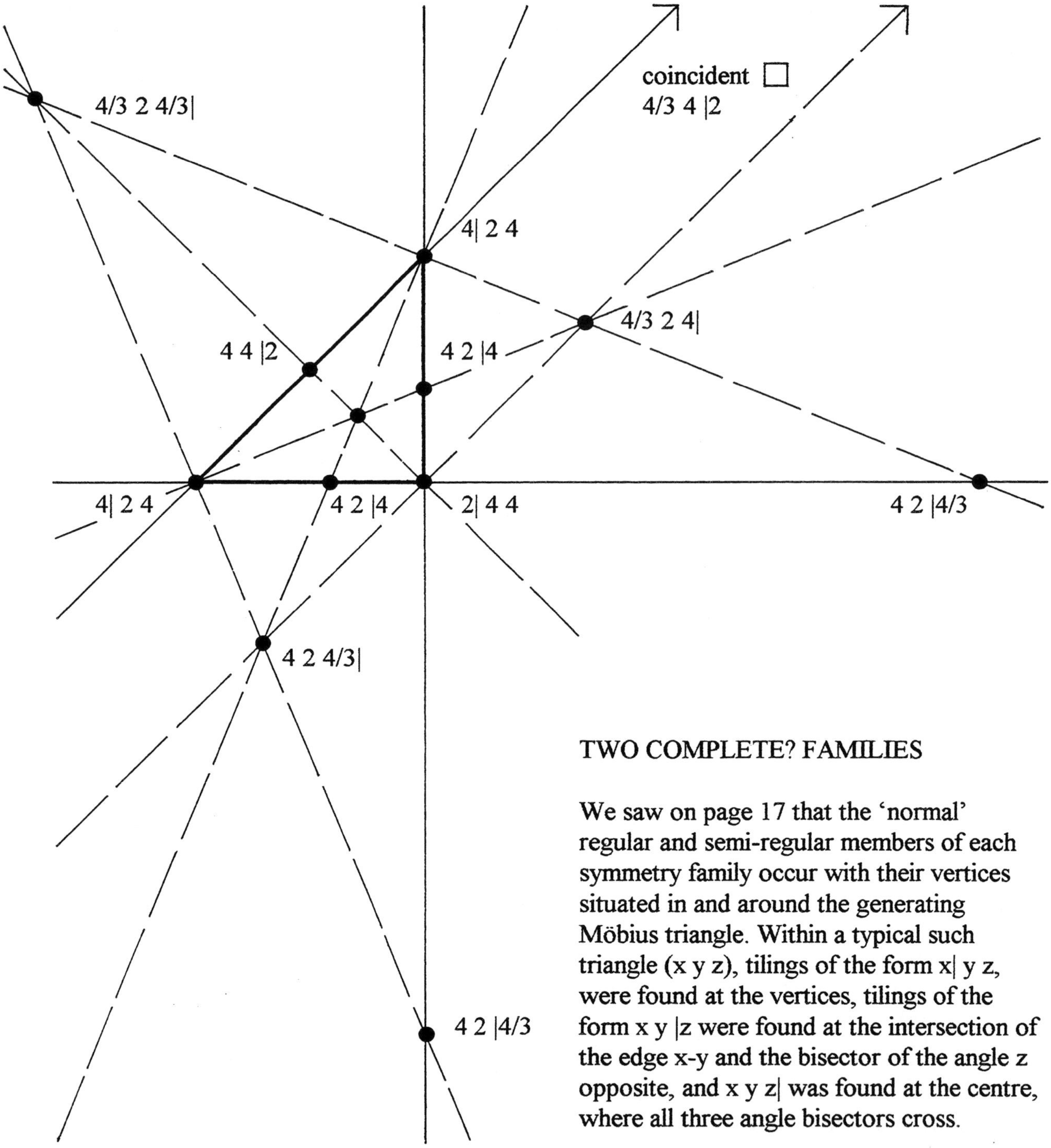

Fig.26: The (4 2 4) Family

TWO COMPLETE? FAMILIES

We saw on page 17 that the 'normal' regular and semi-regular members of each symmetry family occur with their vertices situated in and around the generating Möbius triangle. Within a typical such triangle (x y z), tilings of the form x| y z, were found at the vertices, tilings of the form x y |z were found at the intersection of the edge x-y and the bisector of the angle z opposite, and x y z| was found at the centre, where all three angle bisectors cross.

Within each Möbius triangle the internal angle of a vertex can be given as $\theta/2$, where θ is the external angle of the polygon whose symmetry is represented by that vertex. Thus for a polygon $\{n/d\}$, $\theta = (360d/n)°$, so that the angle at each vertex of the Möbius triangle is $(180d/n)°$.

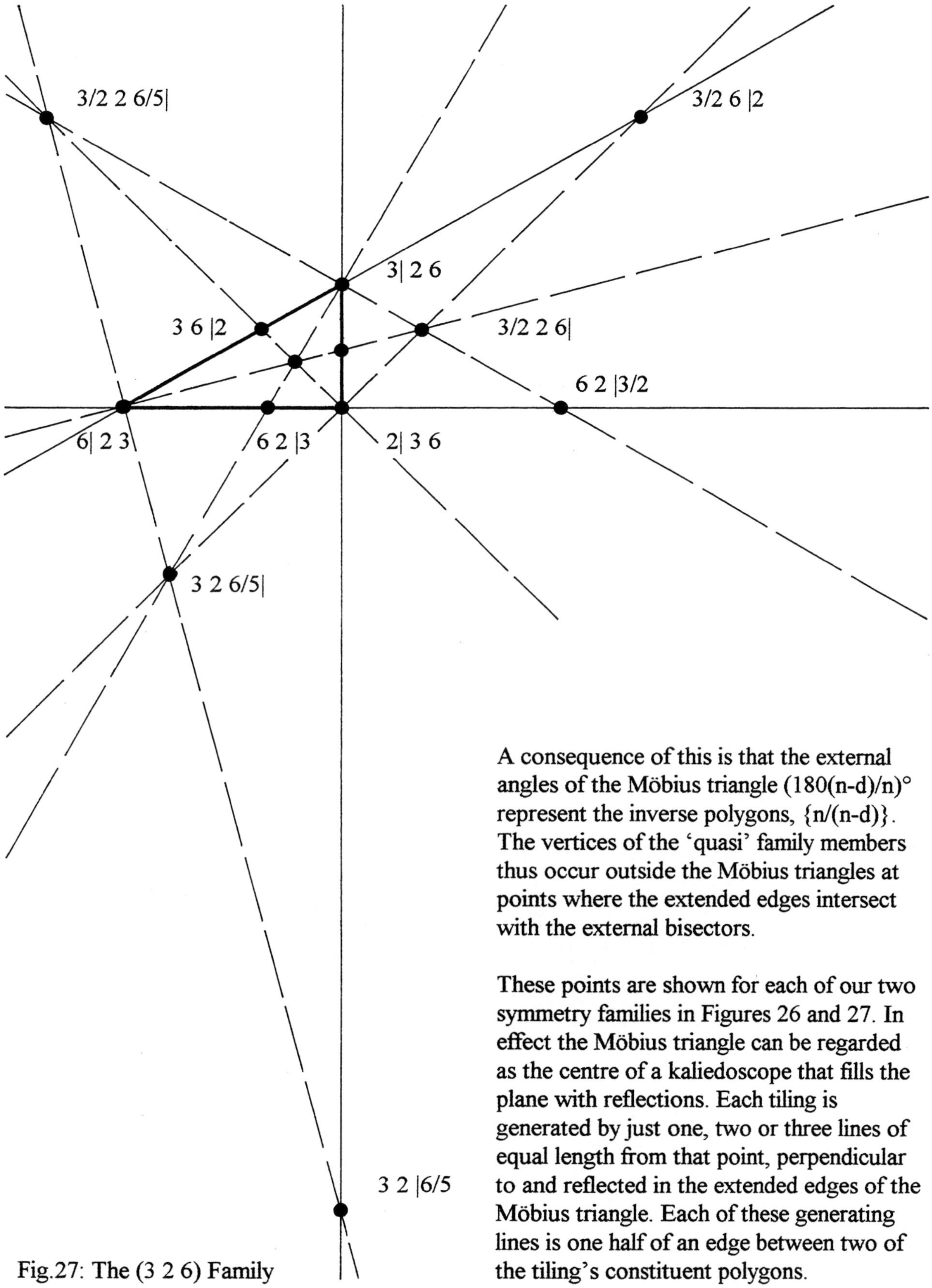

Fig.27: The (3 2 6) Family

A consequence of this is that the external angles of the Möbius triangle (180(n-d)/n)° represent the inverse polygons, {n/(n-d)}. The vertices of the 'quasi' family members thus occur outside the Möbius triangles at points where the extended edges intersect with the external bisectors.

These points are shown for each of our two symmetry families in Figures 26 and 27. In effect the Möbius triangle can be regarded as the centre of a kaliedoscope that fills the plane with reflections. Each tiling is generated by just one, two or three lines of equal length from that point, perpendicular to and reflected in the extended edges of the Möbius triangle. Each of these generating lines is one half of an edge between two of the tiling's constituent polygons.

THE OTHER TILING

Before moving on to more complex matters, we should perhaps for completeness include here the last of Grünbaum & Shephard's 11 semi-regular tilings, where the vertices are all the same. This odd tiling is a slight rearrangement of the snub tiling in the {4,4} family with two squares and three triangles at each vertex, as shown in Figure 28.

Like the snubs it cannot be obtained by truncation, nor does it yield further tilings by its own truncation.

With no affinity to a particular symmetry family it has no special symbol in either Schläfli or Coxeter's notation.

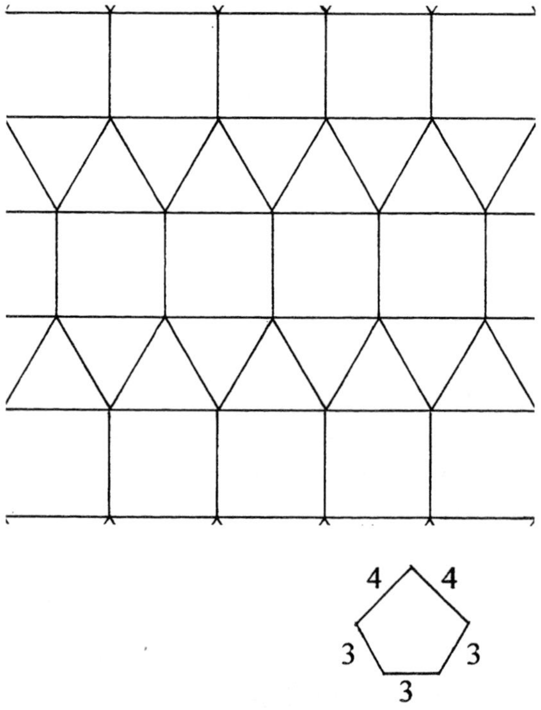

Fig.28: The Other Tiling

END OF PART ONE

Starting with the regular tilings and then truncating and quasitruncating, we have so far explored various ways of assembling regular polygons and their truncations and quasitruncations in the plane. Generally the tilings produced have had a density of 1, and by assigning positive or negative density to different polygons in the quasitruncated versions, the density can be maintained at 1, despite there being multiple layers of polygons present. The only star polygons encountered have been {8/3} and {12/5}, the quasitruncations of {4} and {6}, and the only double polygon has been $\{^{2}3/1\}$, the quasitruncation of {3}.

The second half of this book concerns the star tilings proper and will introduce not only the star polygon {6/2}, along with its truncation $\{^{2}6/1\}$ and quasitruncation $\{^{2}6/2\}$, both double polygons, but also the cross polygons {4/2} and {6/3}, along with their truncations, the double polygon $\{^{2}4/1\}$ and the star polygon {12/3}, all as shown in Figure 29. (It should be remembered that the cross polygons are self inverses with n/d = 2, and therefore do not have quasitruncations).

These polygons are compounds where the 'n' and 'd' of {n/d} are not relatively prime. Figure 30 overleaf shows a table of the polygons from 'The Complete? Polygon', with the compounds conspicuously omitted. Such compounds were originally included as necessary products of the process of forming polygons, without which a classification remains incomplete.

The inclusion of compounds here will allow us to explore a whole new symmetry family of star tilings, the already hinted at {6/2,6}, along with several sub-families of both that family and those already discussed.

Hopefully the value of the apparent heresy of introducing these compound star, cross and double polygons will become clearer as we proceed and eventually be justified by the results produced. Future volumes of this series will apply the same logic to the polyhedra with equally interesting results.

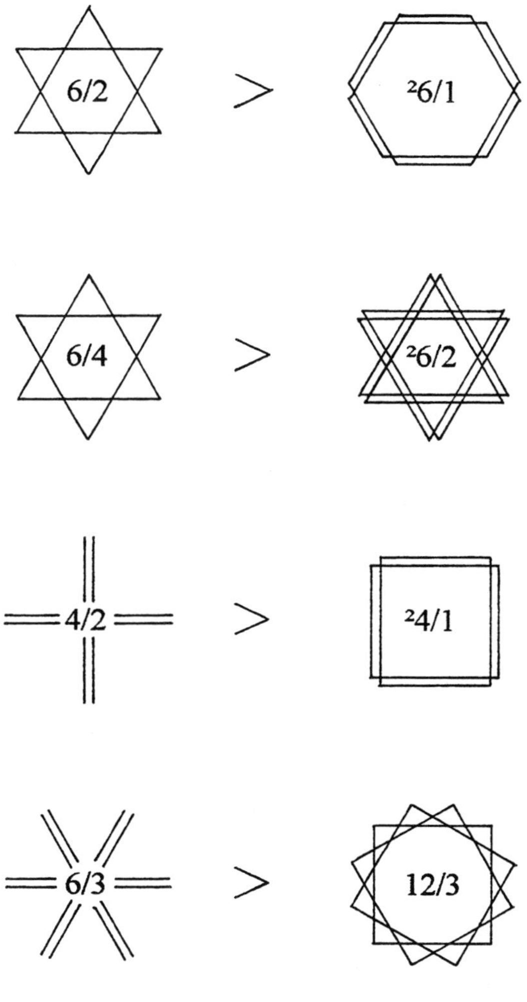

Fig.29: Truncating Compound Polygons

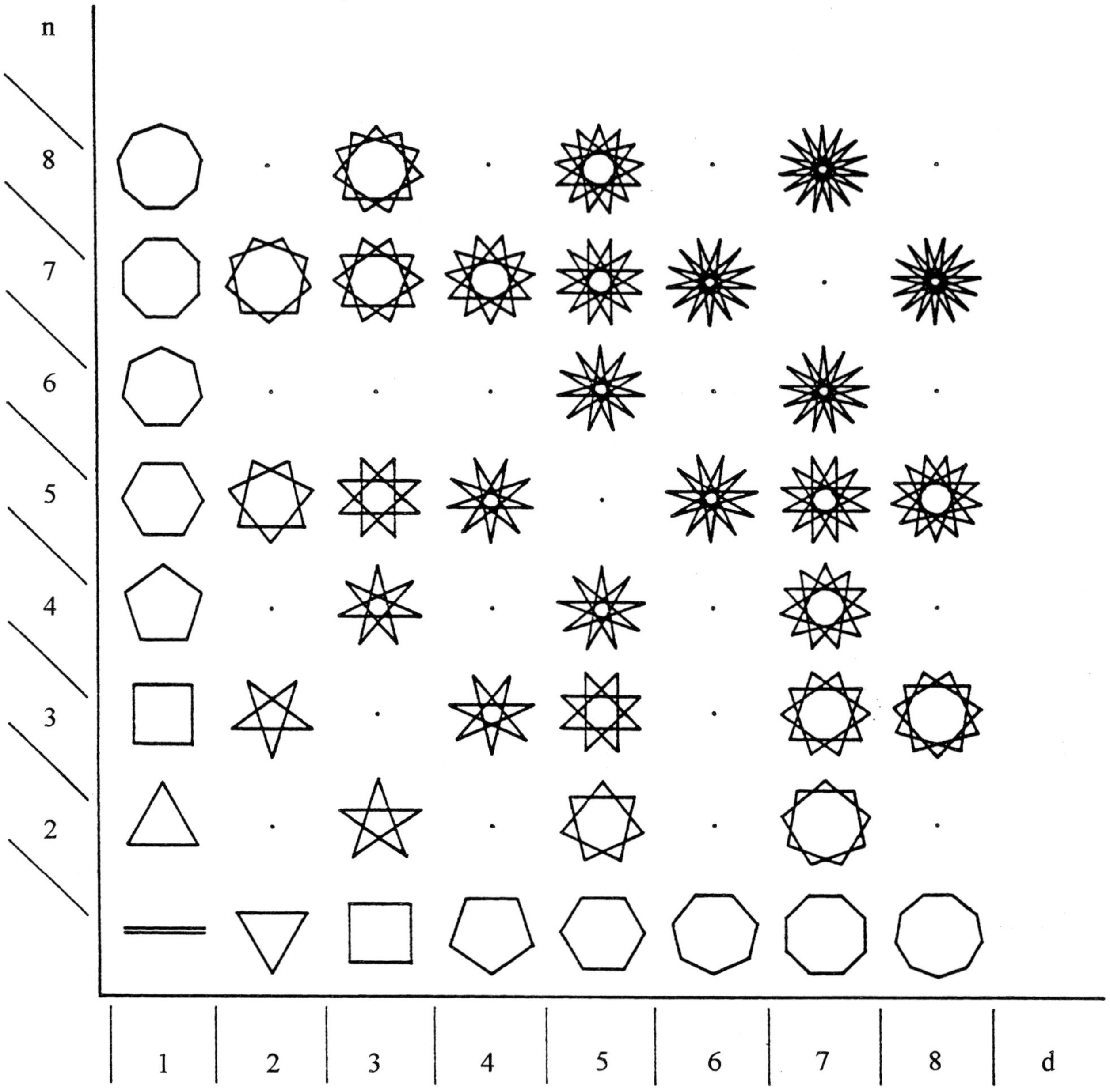

Fig.30: Non-Compound Polygons

SCHWARZ TRIANGLES

We have thus far examined two symmetry families of tilings based on two different grids of Möbius triangles, each a quarter rhomb in the family's respective rhombic tiling. By joining all similar points within the grid we have defined for each family a series of parent regular tilings and derived semi-regular or truncated tilings. We have also looked at quasitruncated tilings of various types which can be considered to have vertices which fall outside the Möbius triangles.

If we examine the Möbius triangle grids we can see various other larger triangles within them that will prove worthy of further investigation, as they too can be repeated by rotation and reflection to fill the entire grid. These are shown in Figure 31. The (4 2 4) family throws out one further possibility in the form of (4 4/2 4), whilst the (3 2 6) family is far more productive with (3 6/2 3), (6 3/2 6), (3 6/3 6), (6/2 2 6), (6 6/4 6), (6/2 6/2 6/2) and (6/2 6/3 6).

These larger triangles are known as Schwarz triangles and hold the key to a number of sub-families and a super-family related to those we have already examined, along with a whole new family of star tilings with its own sub-families and super-family.

The simplest Schwarz triangles consist of two of our Möbius triangles placed back to back. These are the basis of the (4 4/2 4), (3 6/2 3), (6 3/2 6) sub-families and are shown overleaf in Figure 32. These can be generalised as (x y/2 x), allowing us to examine all three sub-families together. The {y/2} produced here occurs at the apex of the two back to back triangles, and is formed from two {y} vertices of the original Möbius triangles.

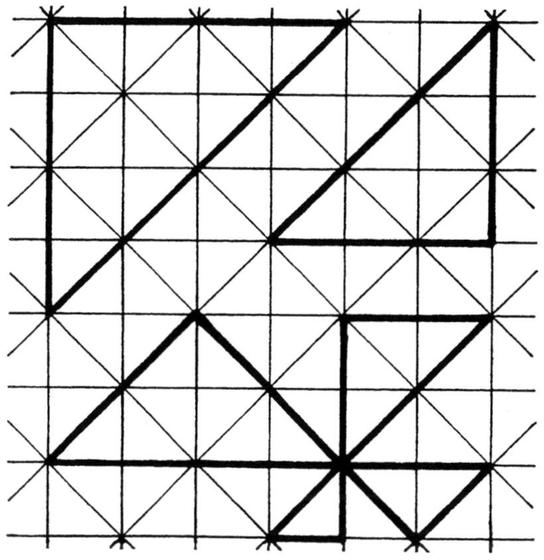

(4 2 4) & (4 4/2 4)

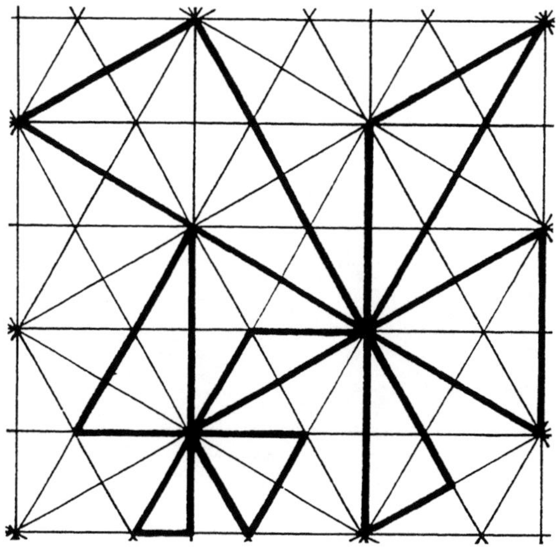

(3 2 6), (3 6/2 3), (6 3/2 6), (3 6/3 6), (6/2 2 6), (6 6/4 6), (6/2 6/2 6/2) & (6/2 6/3 6)

Fig.31: Möbius and Schwarz Triangles

DOUBLE TILINGS

At the apex of each Schwarz triangle we find y/2| x x, which will have a $\{^2y\}$ polygonal vertex figure, all the faces being {x}. This is in fact a doubled up version of y| 2 x or {x,y} one of the original parent regular tilings.

At the centre of each Schwarz triangle we find an even faced tiling x y/2 x|, whose triangular vertex figure will include the polygons t{x}, t{x} and $\{^2y\}$, making it a doubled up version of y 2 |x or t{x,y}, the truncation of the same parent regular tiling.

Elsewhere along the Schwarz triangle's axis of symmetry, this doubling up continues and can be summarised as follows:
y/2| x x = 2(y| 2 x), a double parent tiling
x y/2 x| = 2(y 2 |x), a double truncated tiling
x x |y/2 = 2(2| x y), a double quasi-regular tiling and finally
x' y/2 x'| = 2(y 2 |x'), double quasitruncated tiling.

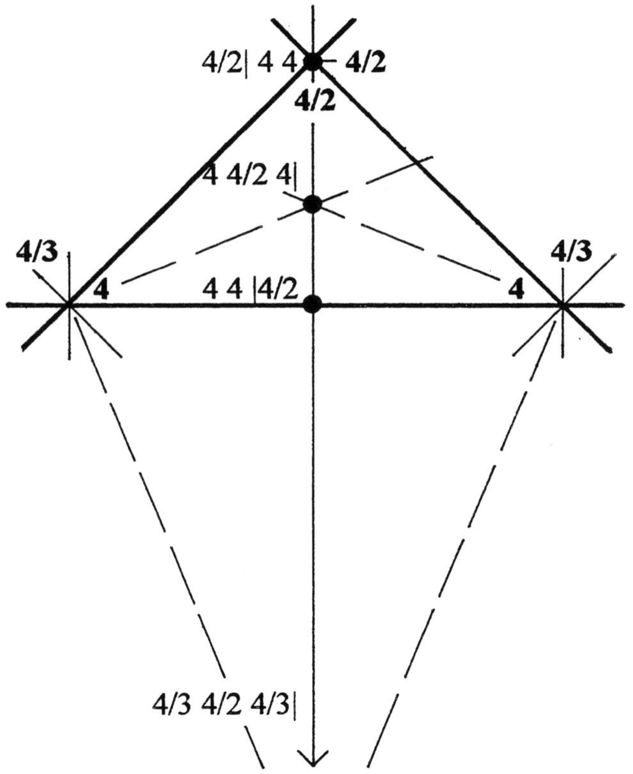

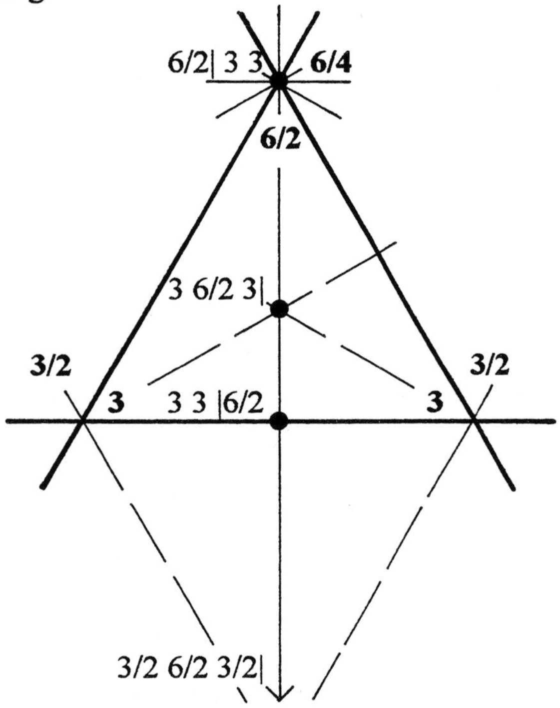

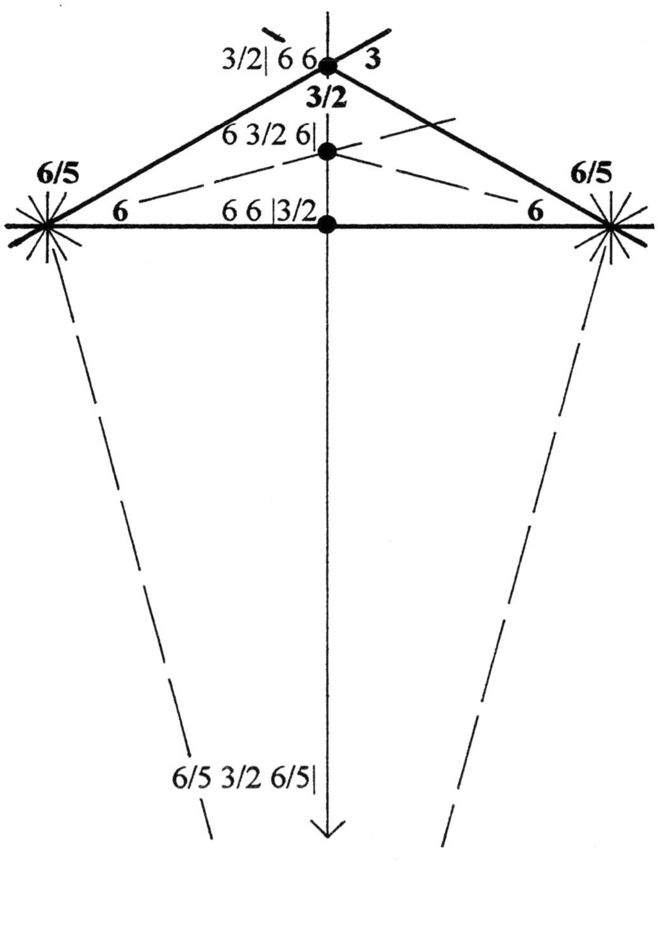

Fig.32: Sub-Family Schwarz Triangles

MORE QUASI-REGULAR TILINGS

From the results so far it might be thought that these new Schwarz triangles are going to yield nothing other than double density versions of the tilings we have already seen. However this is not the case as we will see by considering locations around the Schwarz triangles away from the axis.

The lower vertex of each triangle is in general x| y/2 x, which has a {2x} polygonal vertex figure with faces {x} and {y/2} alternating around it, a kind of quasi-regular tiling. This has effectively two separable layers, each one an original regular tiling, overlaid and joined by the {y/2} faces.

These tilings thus have double density and are shown for each sub-family in Figure 33. They are not quite like anything seen before as the new polygons {6/2} and {4/2} have now at last appeared.

4| 4/2 4

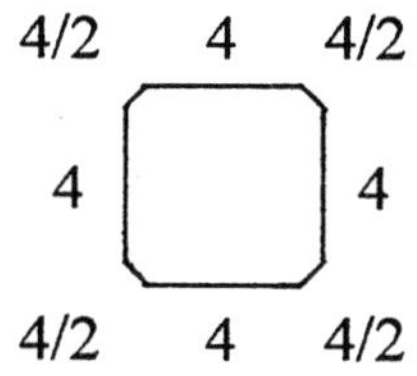

3| 6/2 3

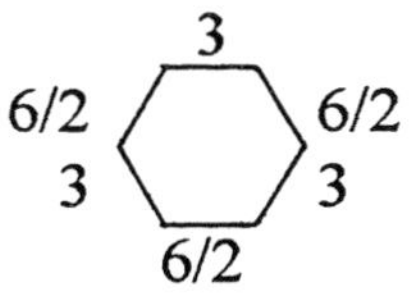

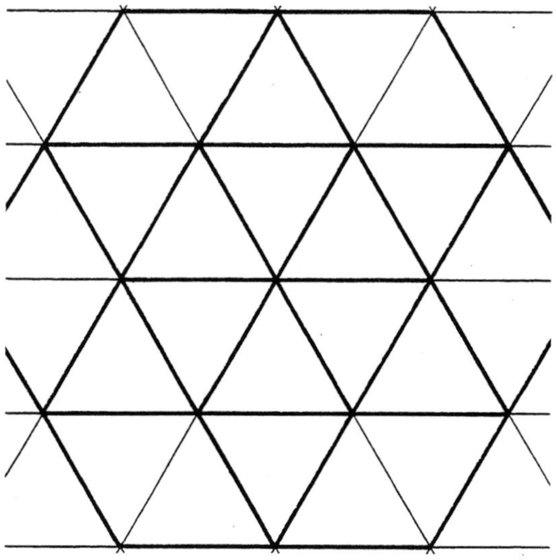

6| 3/2 6

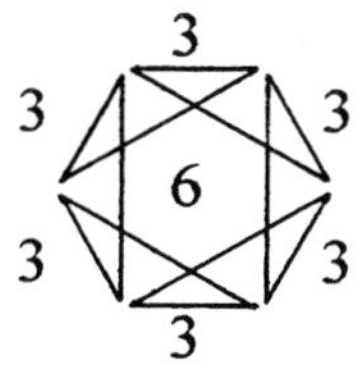

Fig.33: More Quasi-Regular Tilings

MORE RHOMBI TILINGS

Moving now to the upper edges of our Schwarz triangles we have in general x y/2 |x, which can be regarded as the end result of truncating the tilings just described via the central even faced tiling in the triangle. This has a trapezoid vertex figure with faces {x} and {y/2} opposite each other, separated by two faces t{x} = {2x}, a kind of 'rhombi' tiling.

The corresponding tilings are shown for each sub-family in Figure 34, where it can be seen that two of our sub-families produce totally new patterns, the third producing the p-type rhombi tiling of the {3,6} family which was defined earlier on page 27.

These again have double density, with the peculiar property that in each tiling the largest (truncating) polygon {2x} is centred around one of half its value {x}.

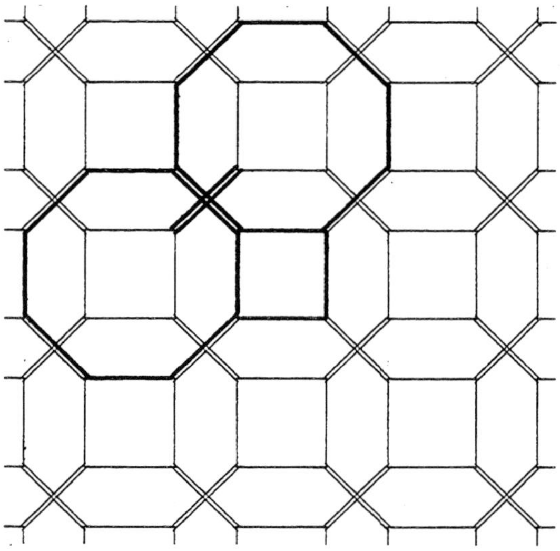

4 4/2 |4

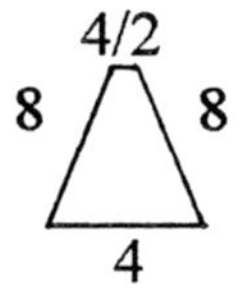

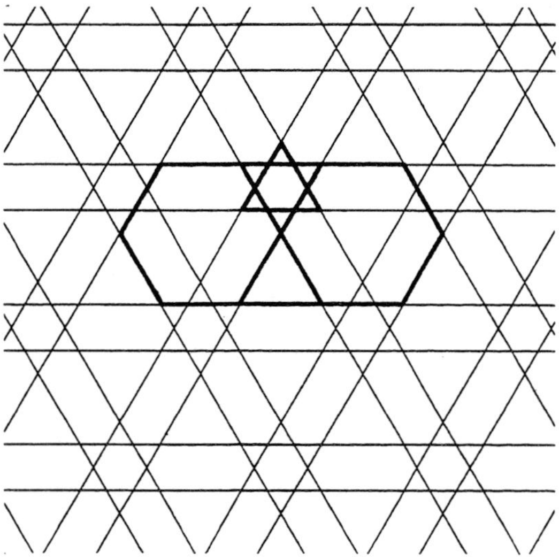

3 6/2 |3

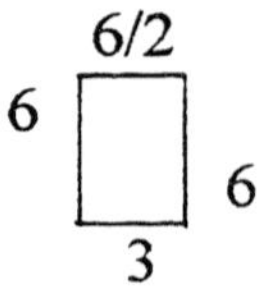

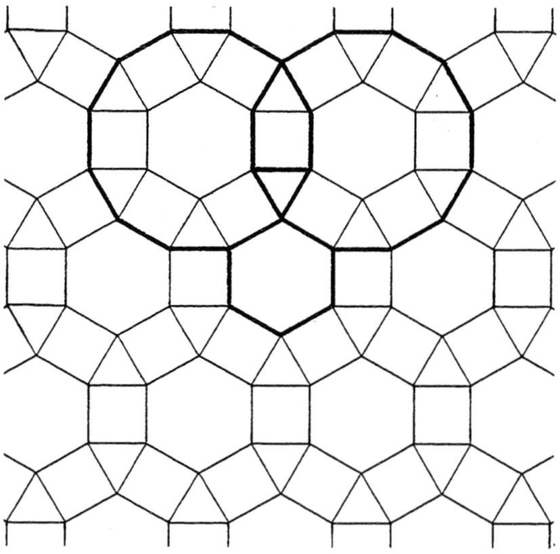

6 3/2 |6 $p\begin{Bmatrix}3\\6\end{Bmatrix}$

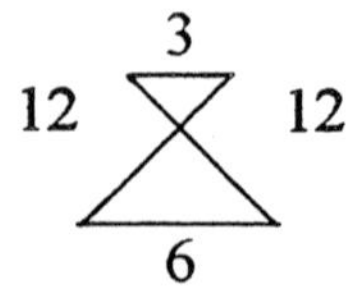

Fig.34: More Rhombi Tilings

MORE EVEN FACED TILINGS

Having now considered all locations on the main Schwarz triangle, we must now look at locations further afield, i.e. quasitruncations. Three even faced possibilities arise in addition to that at the centre of the triangle. The first of these is in the lower triangle on the central axis and has already been dispensed with on page 36; the other two are in the upper left and right triangles and are one and the same on account of the Schwarz triangle's symmetry.

In general this tiling has a triangular vertex figure and is a quasitruncation with symbol x (y/2)' x'|. Its faces are the truncations of these values, the conjugate pair t{x} and t{x'} sharing the same centres. These tilings are shown for each sub-family in Figure 35, where double density can still be observed by once again taking the largest polygon as positive, and the infilling ones at each vertex as negative.

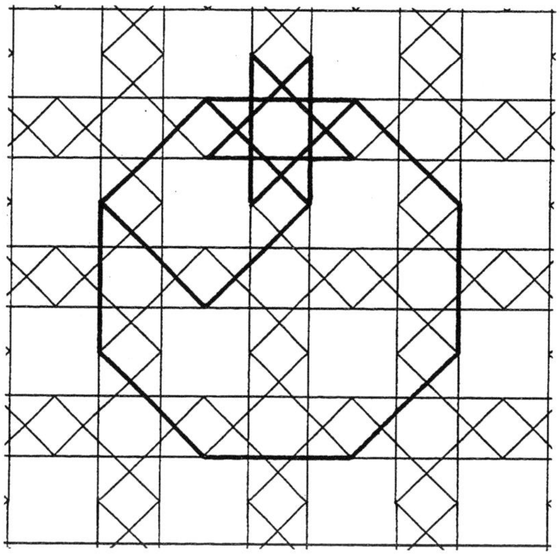

4 4/2 4/3|

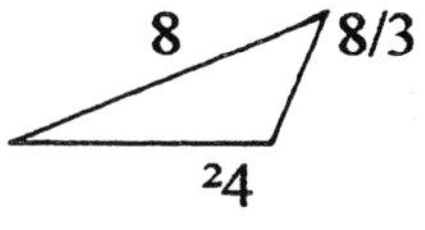

3 6/4 3/2|

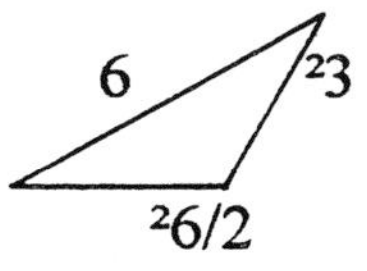

6 3 6/5|

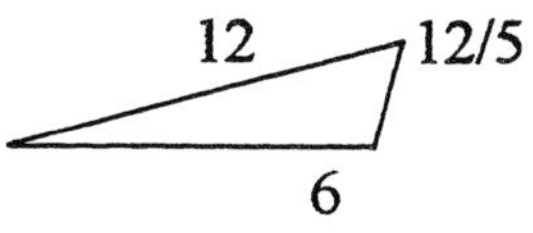

Fig.35: More Even Faced Tilings

MORE QUASIRHOMBI TILINGS

Taking the last set of tilings and continuing the quasitruncation process we end up on the extended upper edge of our Schwarz triangle, at x (y/2)' |x'. These are effectively the 'quasirhombi' members of each sub-family and are illustrated in Figure 36.

Here we find one sub-family produces an entirely new tiling, consisting of two overlaid quasitruncated {4,4} tilings joined by {4/2} faces, whilst a second sub-family produces a coincident {6/2} tiling (qv) and the third the p'-type quasirhombi tiling of the {3,6} family which was also defined earlier on page 27.

The 'rhombi' and 'quasirhombi' forms we have just been examining have no paired squares in their make up and consequently cannot be twisted in the normal way to produce snub forms for these sub-families.

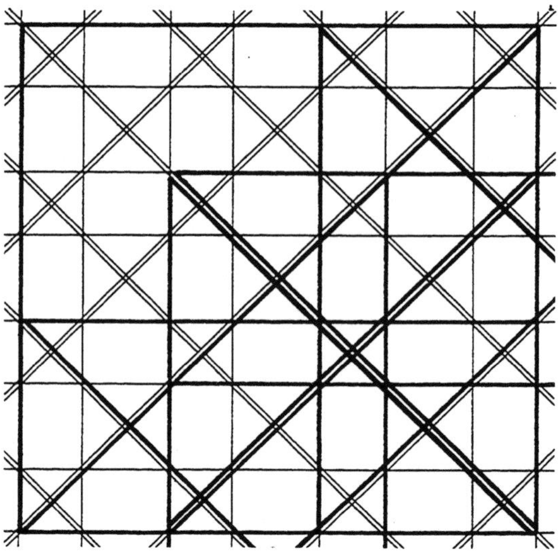

4 4/2 |4/3

4/2

8/3 8/3

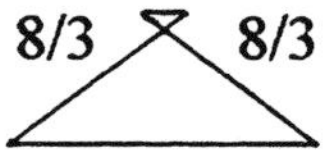

4

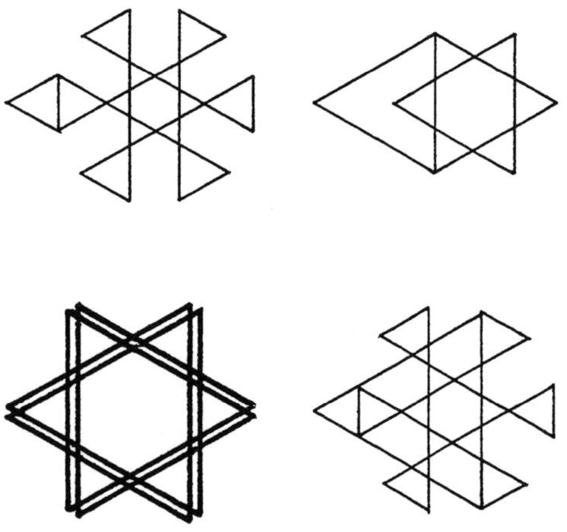

3 6/4 |3/2

6/2

²3 ²3

3

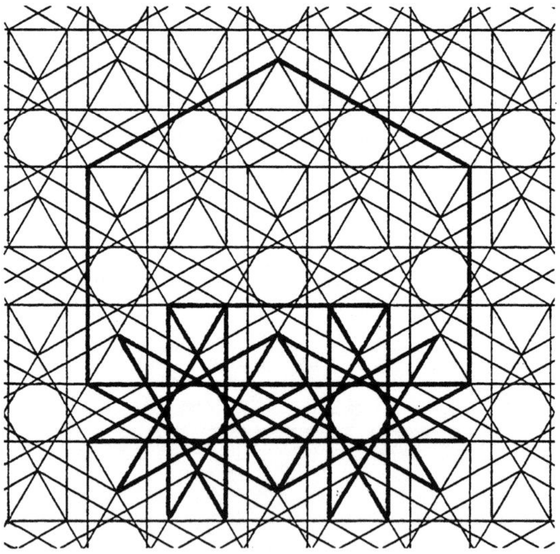

6 3 |6/5 p'{3 / 6}

3

12/5 12/5

6

Fig.36: More Quasirhombi Tilings

MORE MANIPULATION

In Figures 24 & 25, the even faced tilings of our two original symmetry families were considered as members of new short truncation series. The tilings we have just been examining can also be illustrated in this way as shown in Figure 37.

Here we find that they too form similar sets of figures based on the new even faced tilings as the centres of such short truncation series. Again taking quasitruncations of the starting figures, we find further related even faced tilings, that are actually the doubled up versions of the truncated and quasitruncated regular tilings, discussed on page 36.

Thus had we started with these truncated tilings considered as 'even faced' tilings, we would have arrived at some of the members of the sub-families we have just considered, but not those involving {4/2} or {6/2}.

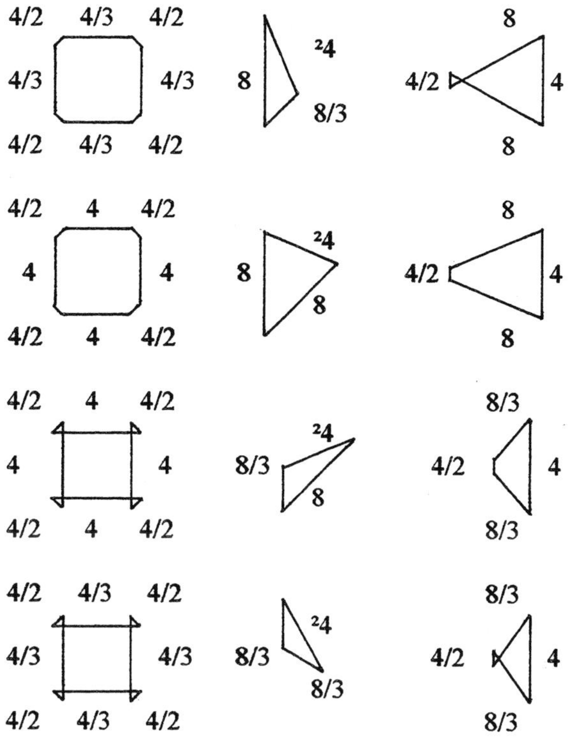

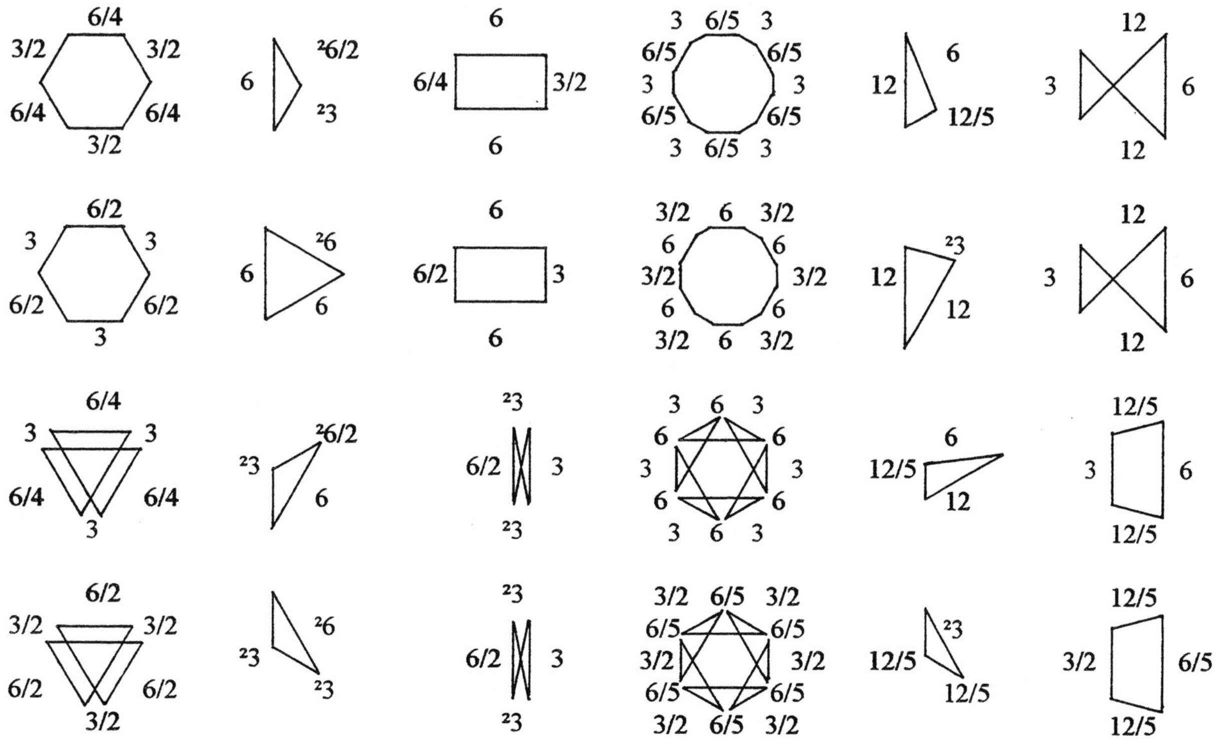

Fig.37: More Short Truncation Series

COINCIDENT TILINGS

In our manipulations on the previous page, one face of our new even faced tilings was not considered as a truncating face. If this is done we find ourselves theoretically on the extended lower edges of our new Schwarz triangles.

The resultant 'rhombi' tilings will generally have the symbol x x' |(y/2)' with a trapezoidal vertex figure consisting of one {x} and one {x'} separated by two t{(y/2)'} faces. These last faces are equiangular with the {x} and {x'} faces so that the trapezoid has no choice other than to fold up into another multiply coincident form.

In terms of the Schwarz triangle the vertices of these coincident tilings are at infinity, since the external bisector at {y/2} is parallel to the edge through the two {x} vertices.

THREE SUB-FAMILIES

Figures 38, 39 and 40 show these coincident tilings within the overall framework of each sub-family. The two coincident tilings previously found can be seen to share a similar property of having vertices at infinity.

In general there are two types of 'quasi-regular' tiling in each sub-family: a double tiling (y/2| x x) and the quasi-regular tiling (x| y/2 x) as shown in Figure 33. There are three 'even faced' tilings: a double truncated parent tiling (x y/2 x|), its quasitruncated equivalent (x' y/2 x'|) and the new even faced tiling (x y/2' x'|) as shown in Figure 35. Finally there are four 'rhombi' type tilings in each sub-family: a double quasi-regular tiling (x x |y/2), the coincident tiling (x x' |y/2'), the rhombi tiling (x y/2 |x) as shown in Figure 34 and the quasirhombi tiling (x y/2' |x') as shown in Figure 36.

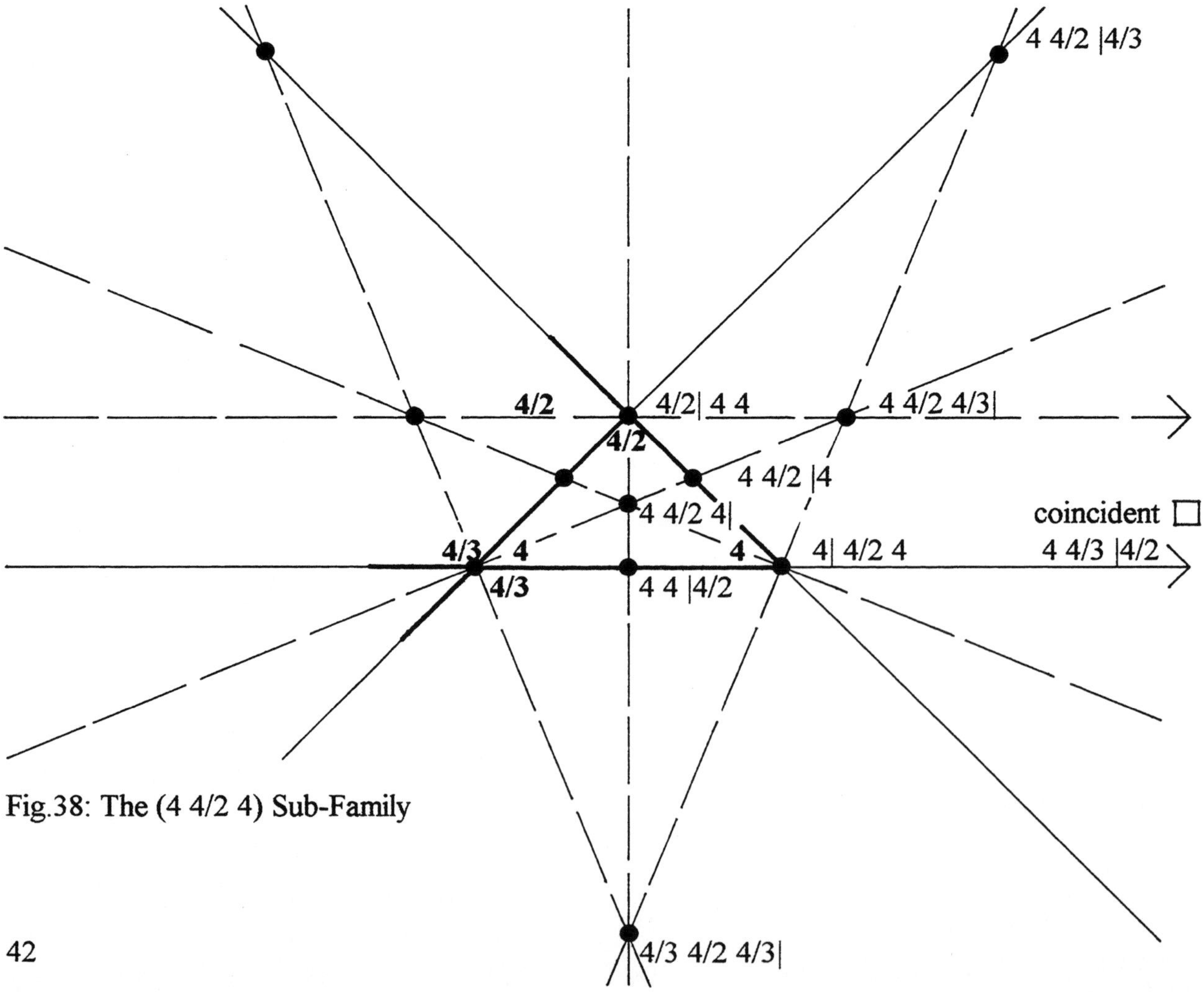

Fig.38: The (4 4/2 4) Sub-Family

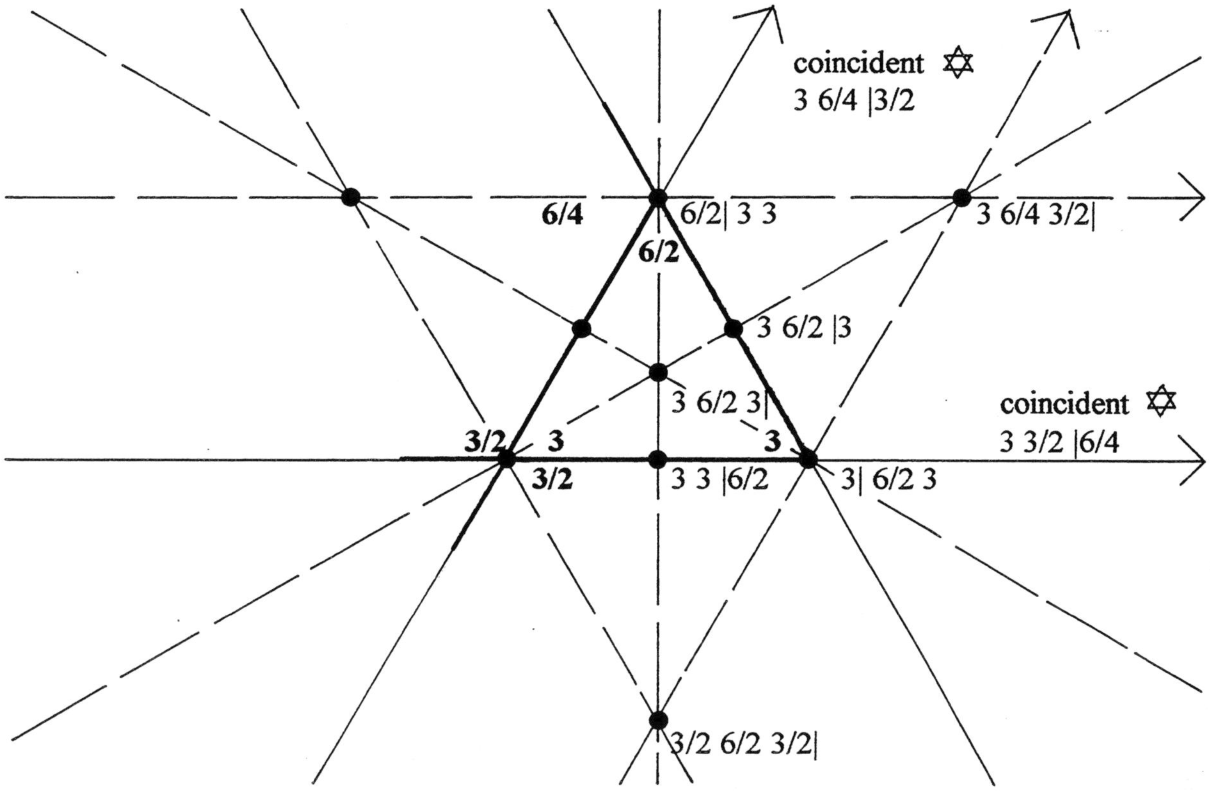

Fig.39: The (3 6/2 3) Sub-Family

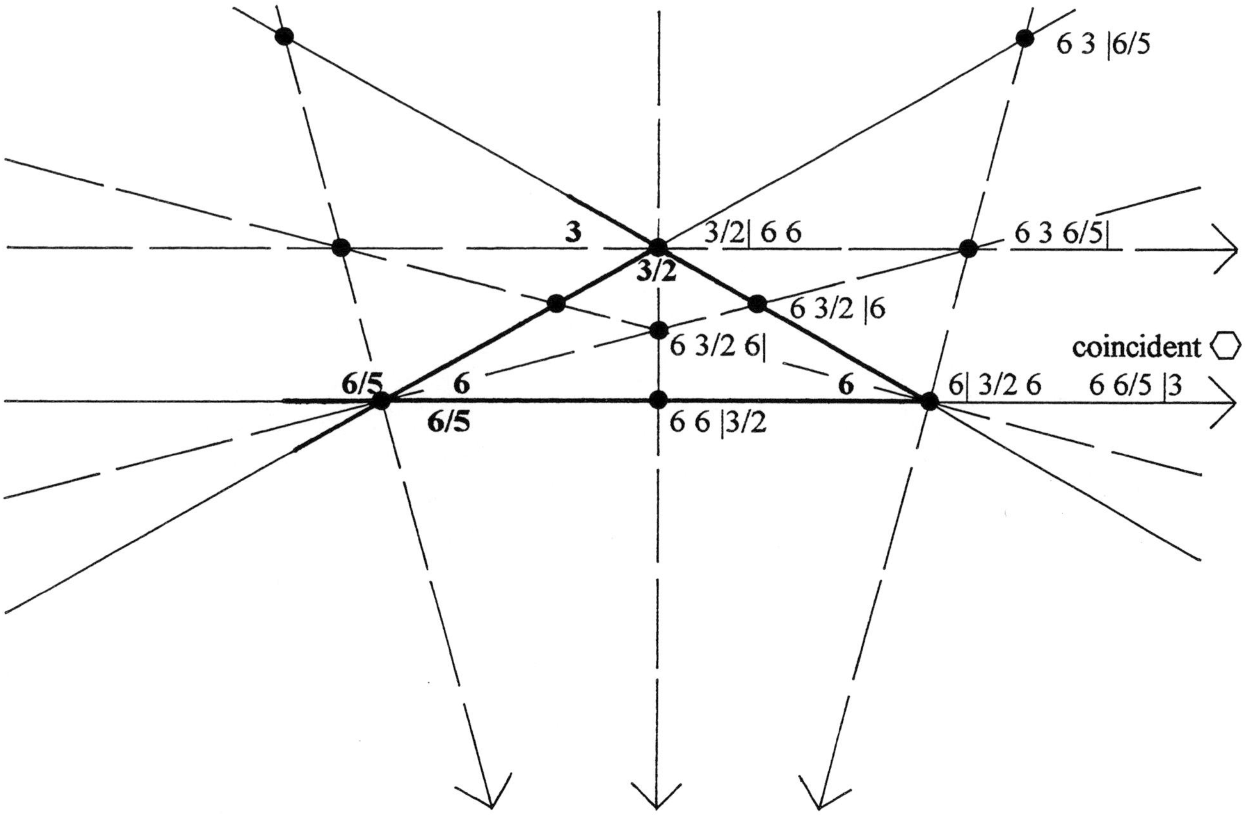

Fig.40: The (6 3/2 6) Sub-Family

6/5 3/2 6/5|

ANOTHER SCHWARZ TRIANGLE

We have just examined the (6 3/2 6) and (3 6/2 3) sub-families, each generated by placing two (3 2 6) Möbius triangles back to back to form their respective Schwarz triangles. A further interesting Schwarz triangle can be formed by placing these two sub-family Schwarz triangles together to produce (3 6/3 6) as shown in Figure 41.

This is very similar to our original (3 2 6) Möbius triangle, but gives rise to a super-family of tilings as a result of the inclusion of the cross polygon {6/3}, and especially its truncation {12/3}.

These are compound polygons which effectively give rise to compound tilings, which will therefore have similarities to the products of the original Möbius triangle. The results, however, make all the effort worthwhile, producing some interesting new patterns in the plane.

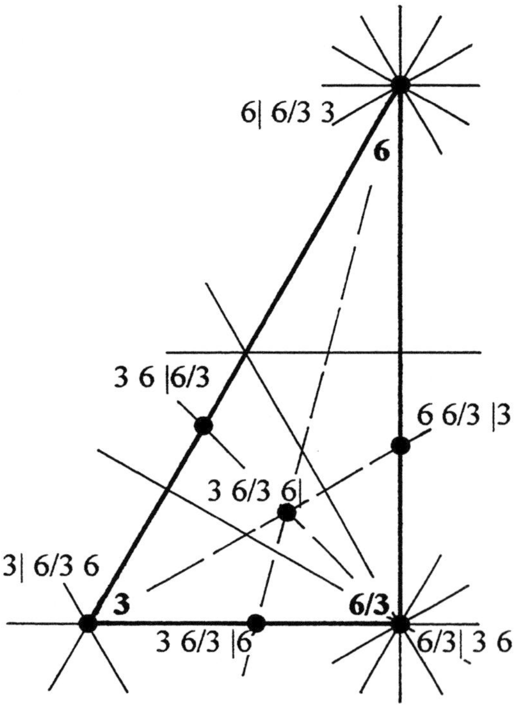

Fig.41: (3 6/3 6) Schwarz Triangle

THREE QUASI-REGULAR TILINGS

Starting with the vertices of our new Schwarz triangle we can generate three new quasi-regular tilings as shown in Figure 42.

These might seem at first sight to be nothing new at all, but we must remember that we are dealing here with a Schwarz triangle consisting of four Möbius triangles, and that in the light of previous experience this is likely to produce fourfold density.

This is in fact the case: 6/3| 3 6 appears to be a simple triangle tiling but actually consists of four overlaid 2| 3 6 quasi-regular tilings, each triangle also being part of three overlapping hexagons, giving a total of four layers. This may seem at odds with the vertex figure which seems to indicate three such layers, however at each vertex we have three pairs of opposite triangles as indicated as well as an overlying hexagon from the fourth layer centred on the vertex.

6| 6/3 3 also appears to be a simple triangle tiling, but here the apparent triangles are only one quarter of the actual face. Each of these small triangles is actually the central cell of one such larger triangle, as well as the overlapping portion of three other larger triangles. The tiling consists of four overlaid 6| 2 3 triangle tilings, with {6/3} faces at their edges holding it all together.

3| 6/3 6 similarly consists of four overlaid 3| 2 6 hexagon tilings held together by {6/3} faces, and looks like the (3 2 6) family's rhombic tiling. Each rhomb, however is the result of two hexagons overlapping and is also covered by a further two overlapped hexagons at right angles to these.

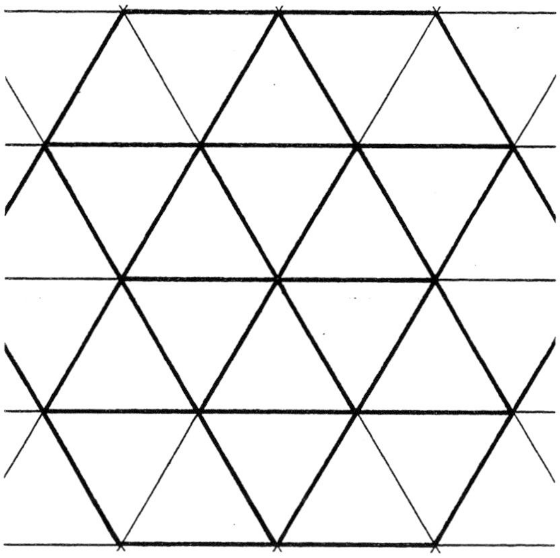

6/3| 3 6

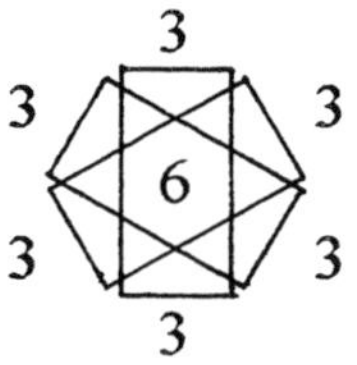

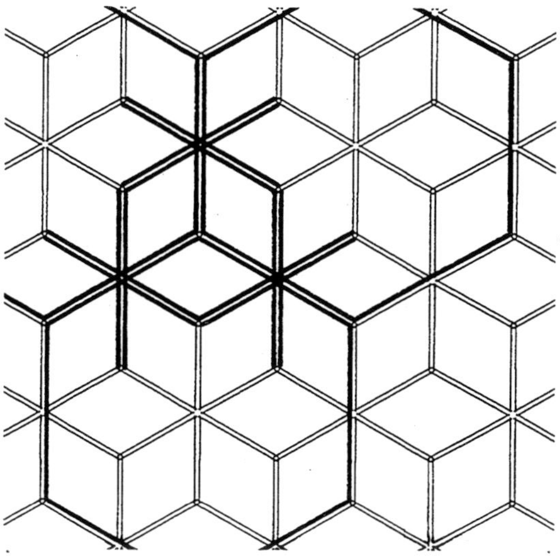

3| 6/3 6

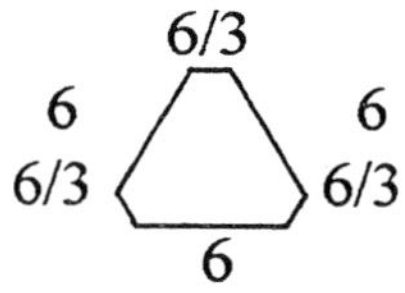

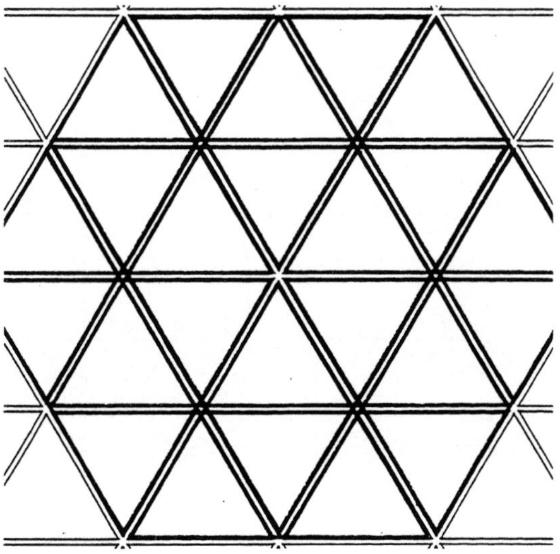

6| 6/3 3

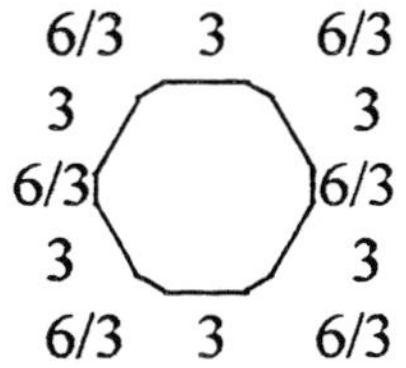

Fig.42: Three Quasi-Regular Tilings

FOUR EVEN FACED TILINGS

If we now move on to the centre of our new Schwarz triangle, and to the centres of the imaginary outlying triangles with inverse values, we can generate four new even faced tilings as shown overleaf in Figure 43.

These exactly follow the (3 2 6) family's even faced tilings with a {12/3} face everywhere in lieu of {4}, so that effectively we have a normal even faced tiling, two quasitruncated even faced tilings and one quasiquasitruncated even faced tiling.

They all have a density of four, perhaps most easily seen in the normal version 3 6/3 6| where each {12/3} of density three is overlain by a concentric {12} with density one. Each tiling is thus a fourfold overlay of the equivalent (3 2 6) family tiling.

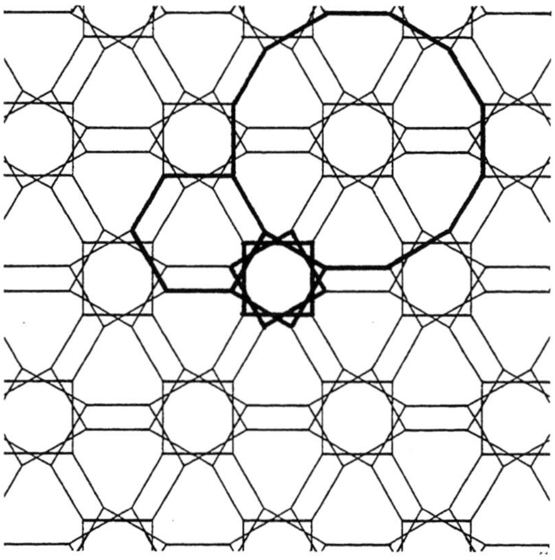

3 6/3 6|

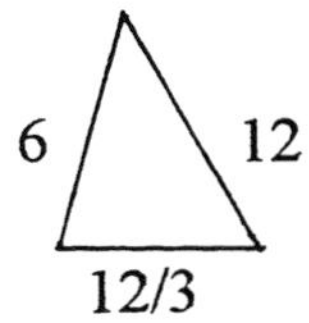

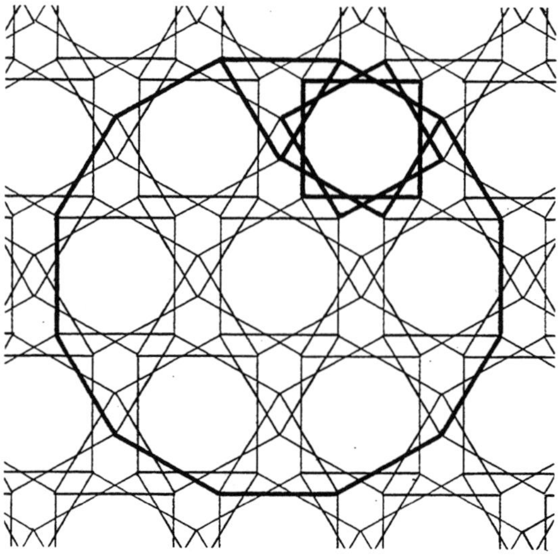

3/2 6/3 6|

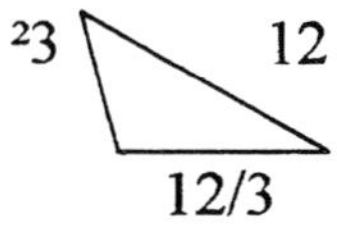

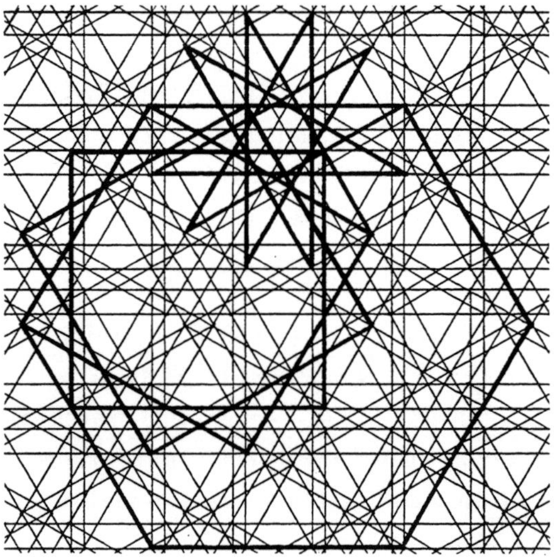

3 6/3 6/5|

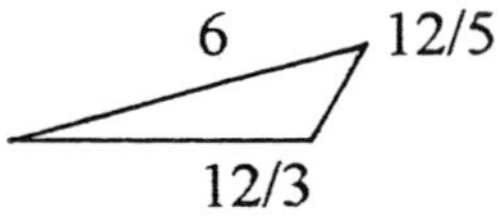

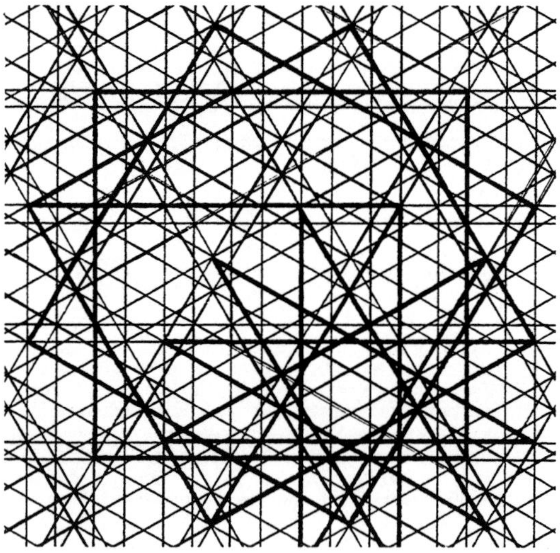

3/2 6/3 6/5|

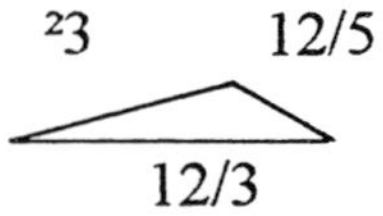

Fig.43: Four Even Faced Tilings

SIX RHOMBI TILINGS

Finally we come to look at the edges of our new Schwarz triangle, along with the extended edges that represent quasi forms.

On the hypotenuse we find 3 6 |6/3 which not surprisingly turns out to be four overlaid 3 6 |2 rhombi tilings. Here the density three {12/3} is concentric with an overlaid {6}, making a total of four. Although the pattern is the same as the previously considered even faced tiling 3/2 6/3 6| it actually contains no {12} faces, providing another parallel with the 3 6 |2 rhombi tiling and its lookalike 3/2 2 6| in the (3 2 6) family.

A quasirhombi tiling can be found to accompany the rhombi tiling in the form of 3/2 6 |6/3 which is, as might be expected, four overlaid 3/2 6 |2 quasirhombi tilings from the (3 2 6) family. Again the pattern is the same as one of the even faced tilings, 3/2 6/3 6/5|, but this time there are no {12/5} faces.

These new rhombi and quasirhombi tilings are shown in Figure 44.

Overleaf in Figure 45 the products of the other edges of the Schwarz triangle are illustrated along with their quasitruncated forms. These are all fourfold versions of the truncated and quasitruncated parent regular tilings originally shown in Figures 3 and 17. Here they no longer have the null {2} face considered on page 29, but instead have an equivalent {6/3}cross polygon as the now familiar glue which knits the layers together.

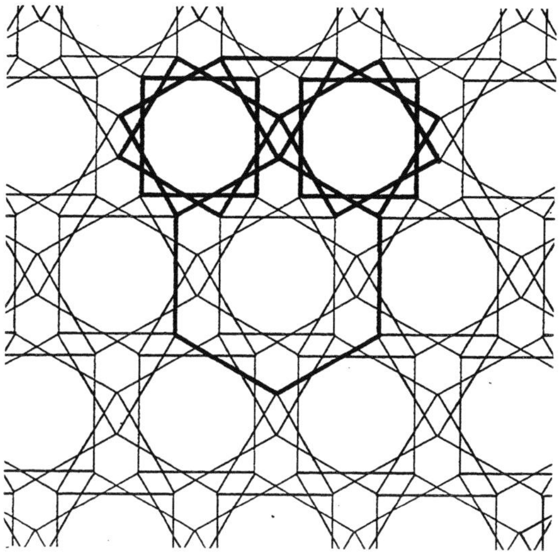

3 6 |6/3

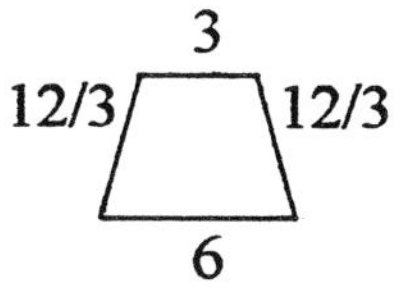

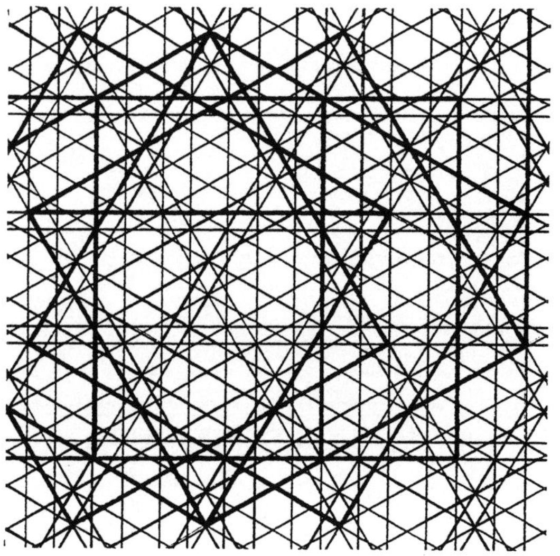

3/2 6 |6/3

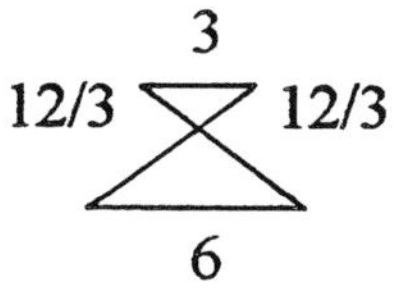

Fig.44: Rhombi and Quasirhombi Tilings

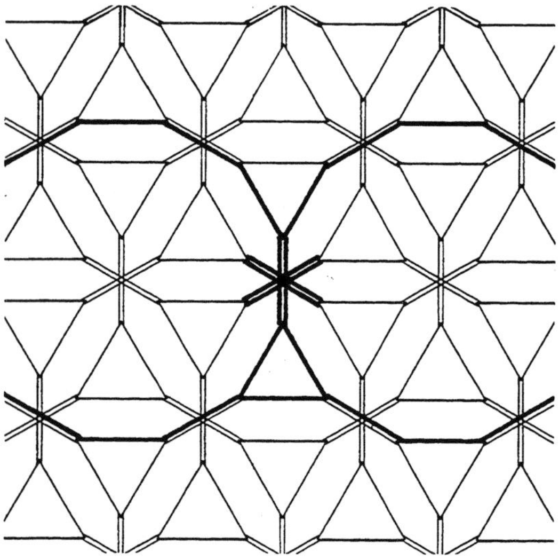

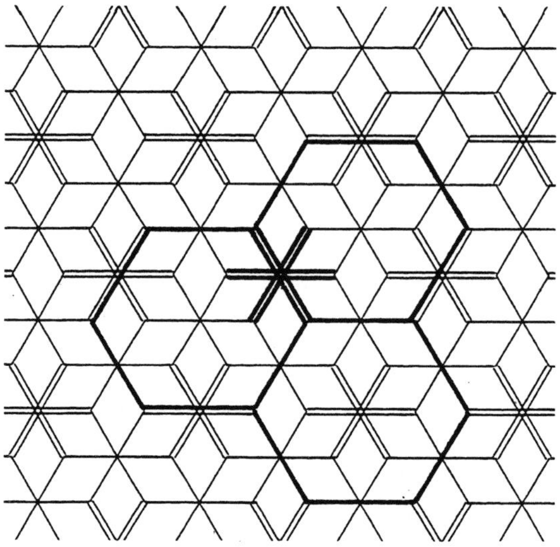

3 6/3 |6

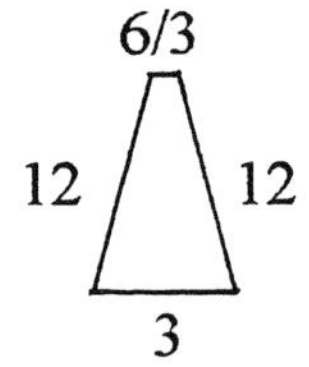

6 6/3 |3

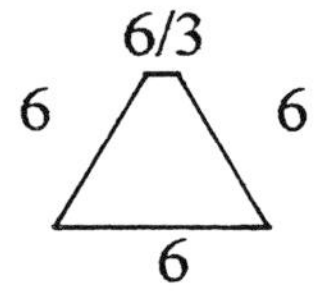

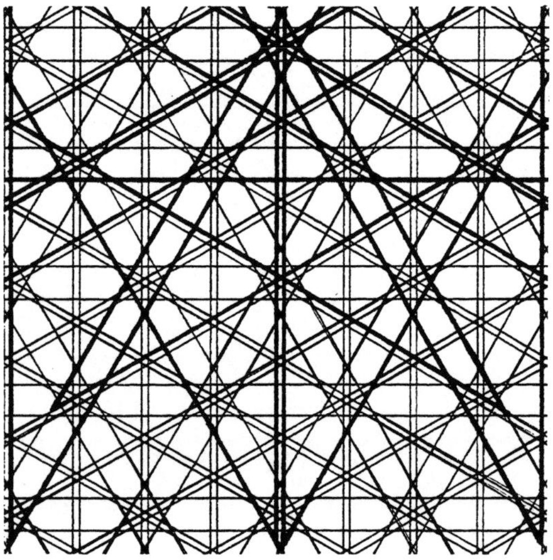

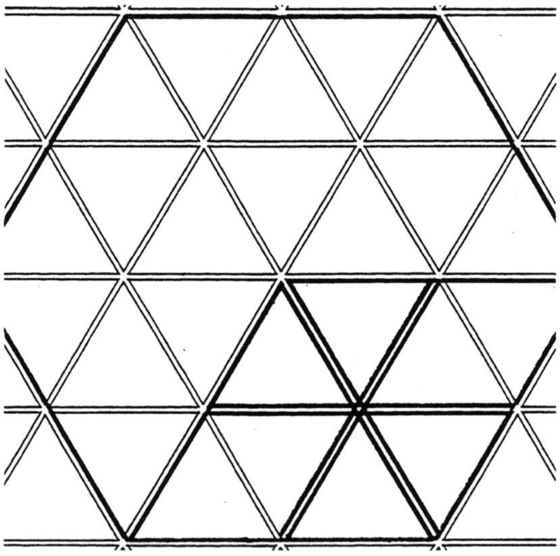

3 6/3 |6/5

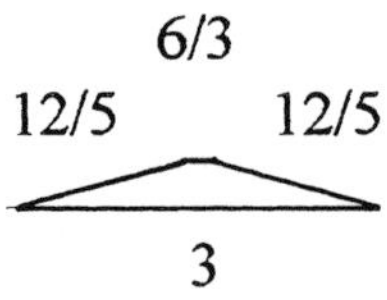

6 6/3 |3/2

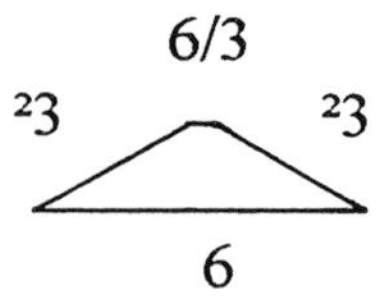

Fig.45: Truncated & Quasitruncated Tilings

TWO SNUB TILINGS

We seem to have found in the (3 6/3 6) super-family a full complement of tilings that parallel those in the (3 2 6) family, as shown overleaf in figure 47; everywhere that a {2} appears it is replaced by {6/3}, a {4} by {12/3}. The obvious question arises as to whether there are equivalent snub tilings in the new super-family.

Whilst a {12/3} is not a {4}, it can be treated as a compound of three {4}s, which if twisted in the usual snub manner become three paired triangles as shown below. Interestingly this produces a {6/3} at their centre!

There are thus both snub and quasisnub tilings completing our new super-family as shown in Figure 46. As before they comprise four overlaid examples of the simpler (3 2 6) family snub and quasisnub tilings.

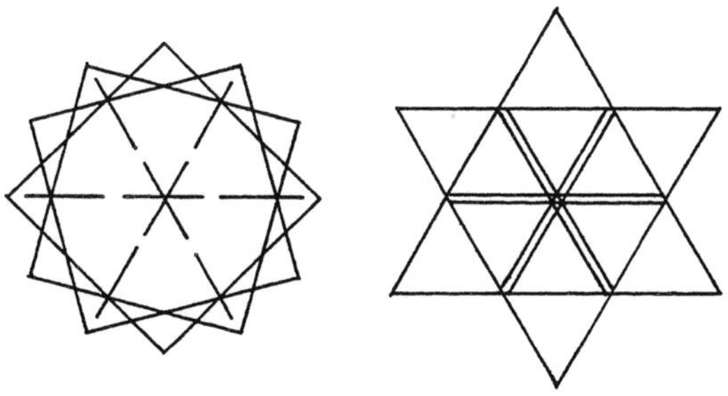

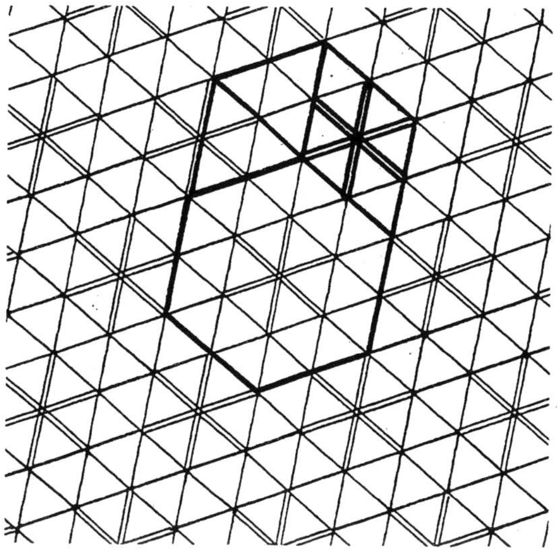

|3 6/3 6

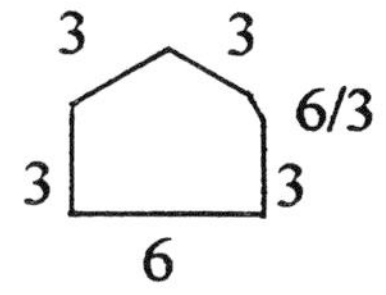

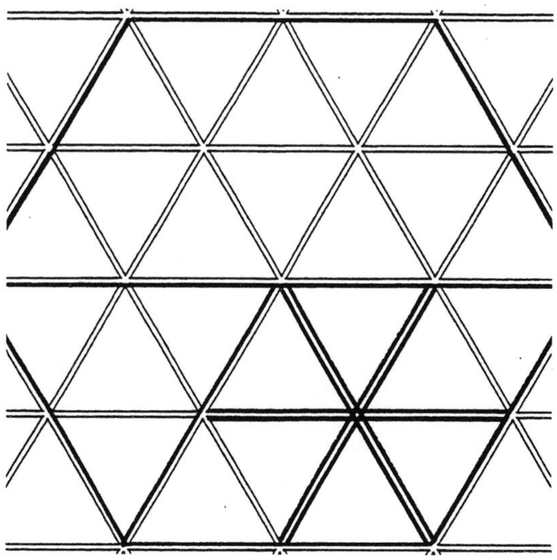

|3/2 6/3 6

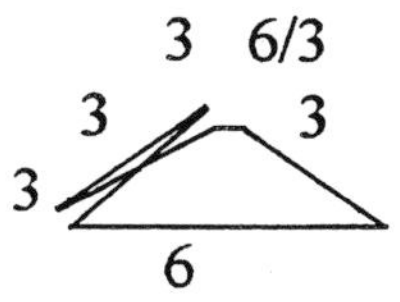

Fig.46: Snub and Quasisnub Tilings

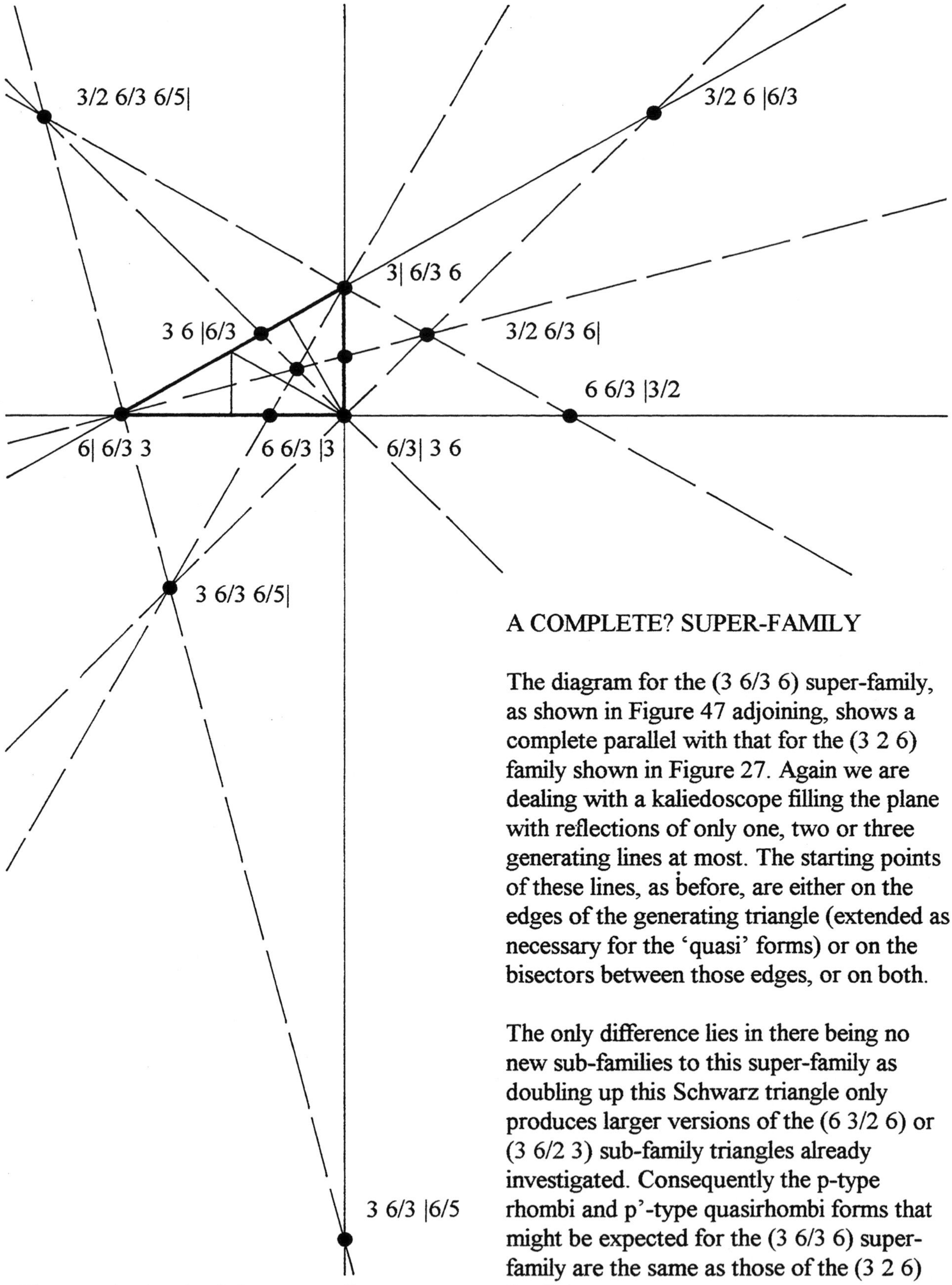

Fig.47: The (3 6/3 6) Super-Family

A COMPLETE? SUPER-FAMILY

The diagram for the (3 6/3 6) super-family, as shown in Figure 47 adjoining, shows a complete parallel with that for the (3 2 6) family shown in Figure 27. Again we are dealing with a kaliedoscope filling the plane with reflections of only one, two or three generating lines at most. The starting points of these lines, as before, are either on the edges of the generating triangle (extended as necessary for the 'quasi' forms) or on the bisectors between those edges, or on both.

The only difference lies in there being no new sub-families to this super-family as doubling up this Schwarz triangle only produces larger versions of the (6 3/2 6) or (3 6/2 3) sub-family triangles already investigated. Consequently the p-type rhombi and p'-type quasirhombi forms that might be expected for the (3 6/3 6) super-family are the same as those of the (3 2 6) family.

THE STAR TILINGS

We now come to examine a further Schwarz triangle, that produced by assembling three of our original Möbius triangles to form (6/2 2 6), as shown in Figure 48.

Like (3 2 6) and (4 2 4), our original regular tiling families, this includes a 2 in its make up, and can thus be considered as a full family of star tilings with equivalent Schläfli notation {6/2,6}. For ease of comparison with the {3,6} family we will continue with this convenient notation here as well as the Coxeter notation that refers back to vertex positions in the (6/2 2 6) Schwarz triangle.

The two parent star tilings are shown in Figure 49 and again at first sight these both appear to be just triangle tilings. As might be expected these actually have a density of three, {6/2,6} consisting of three {3,6} triangle tilings overlaid, whilst {6,6/2} is three {6,3} hexagon tilings overlaid.

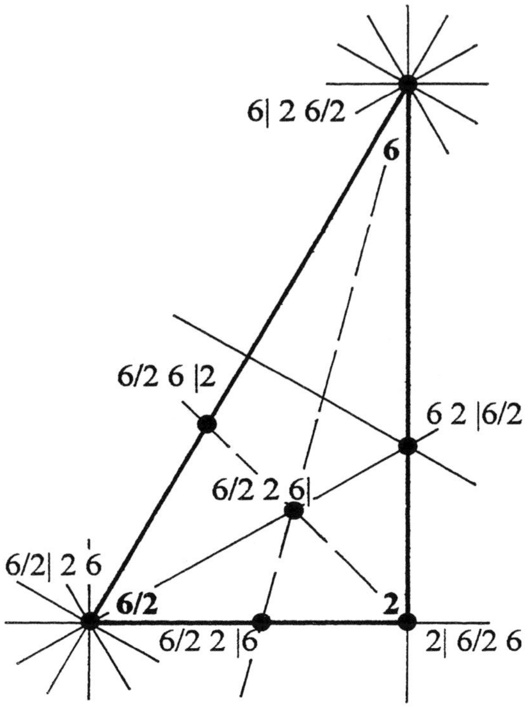

Fig.48: (6/2 2 6) Schwarz Triangle

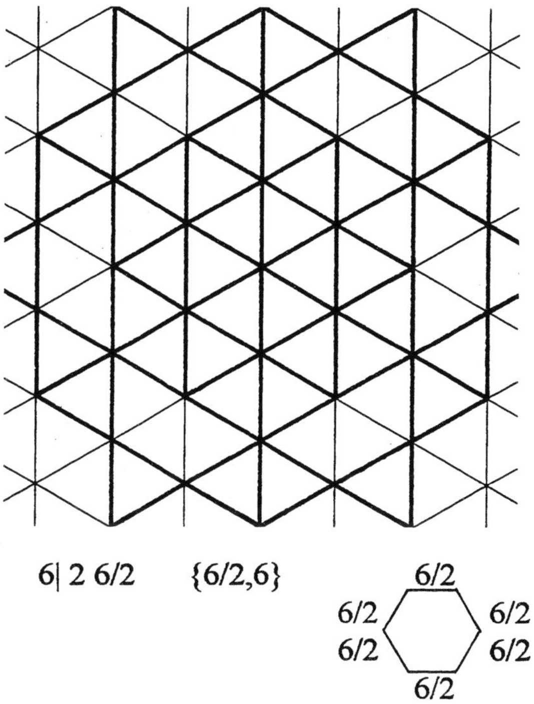

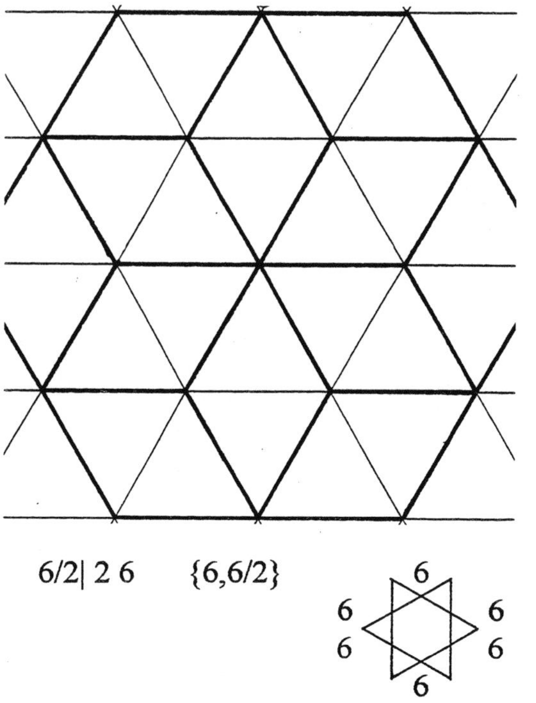

Fig.49: Star Tilings

TRUNCATED STAR TILINGS

The two tilings we have just examined are in fact a dual pair and we can actually form a 'truncation' series between them.

The truncated tilings thus produced again have a density of three and are shown below in Figure 50.

The t{6/2,6} appears to be a hexagon tiling, consisting entirely of hexagons, each of which is actually a single and a double hexagon overlaid (a truncating and a truncated face respectively).

The t{6,6/2} is more interesting, each {6/2} of double density being concentric with an overlying {12} of single density, each area inbetween being covered by three {12}s. These two tilings can also be regarded as three overlaid t{3,6} or t{6,3} respectively.

A QUASI-REGULAR STAR TILING

This first truncation series we are examining should have a central quasi-regular tiling whose vertex figure will be a rectangle with two opposing {6} faces and two opposing {6/2} faces.

This is in fact the case and the resultant quasi-regular star tiling is shown in Figure 51 opposite. It again has a density of three, here each {6/2} of double density being concentric with a {6} of single density.

Whilst this tiling looks like a simple 2| 3 6 quasi-regular tiling, its edges are actually three times longer than they appear.

It can also be seen as three overlaid 2| 3 6 quasi-regular tilings, again maintaining the parallels between the two generating triangles.

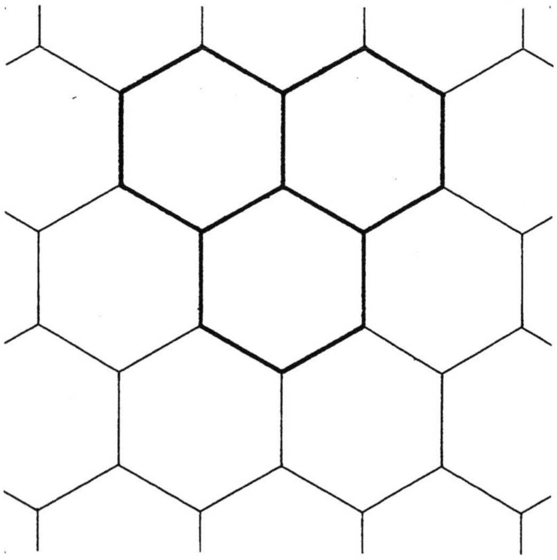

6 2 |6/2 t{6/2,6}

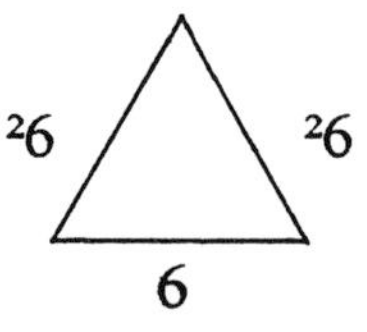

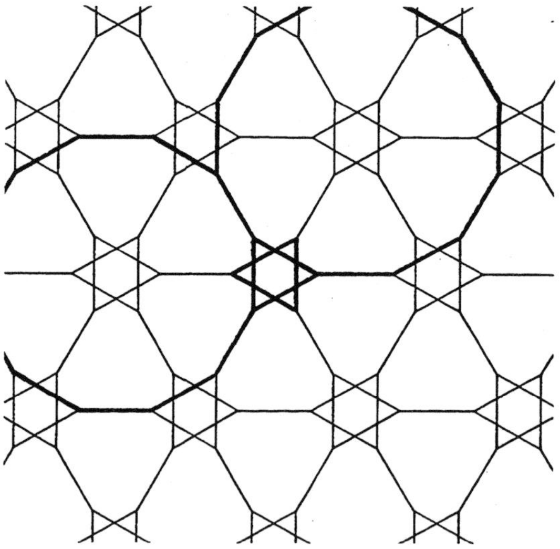

6/2 2 |6 t{6,6/2}

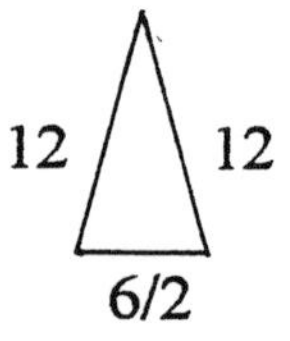

Fig.50: Truncated Star Tilings

QUASITRUNCATED STAR TILINGS

The tilings so far produced for the (6/2 2 6) star tiling family have all been equivalent to three overlaid members of the original (3 2 6) family and have come from equivalent positions on their respective generating triangles.

The parallels are in fact complete, and we can also take either of the parent tilings in our new family and quasitruncate them to produce quasitruncated star tilings that are three overlaid quasitruncated tilings from the original family, as shown in Figure 52. These again have a density of three when the usual rules are applied for quasitruncated tilings: positive density for the largest polygon at each vertex, negative for the infilling ones.

Here we should note that a quasitruncated {6/2} is a truncated {6/4}, i.e. {^{2}6/2}, another parallel to a quasitruncated {3/1} being a truncated {3/2}, i.e. {^{2}3/1}.

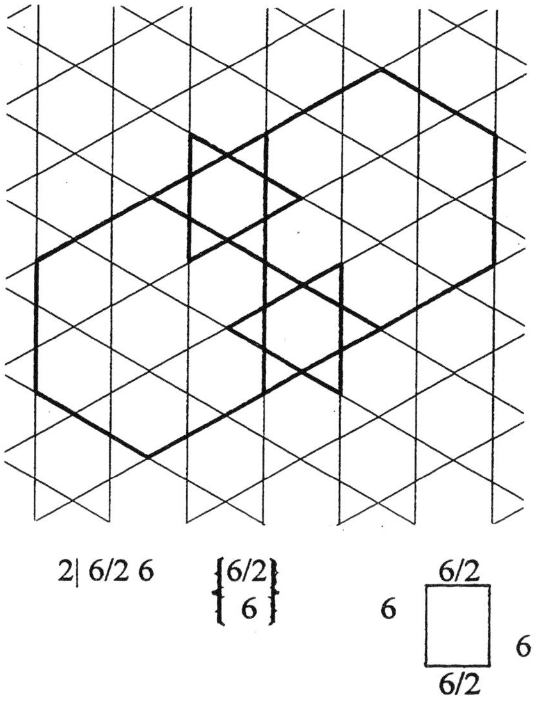

Fig.51: A Quasi-Regular Star Tiling

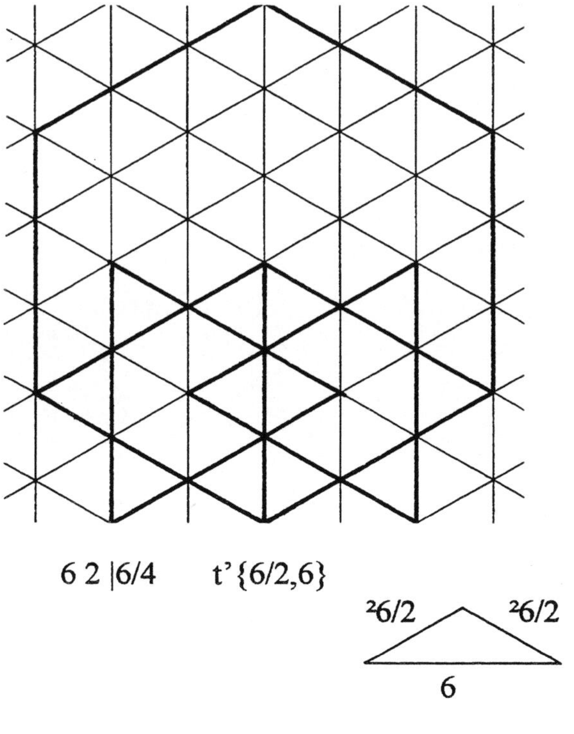

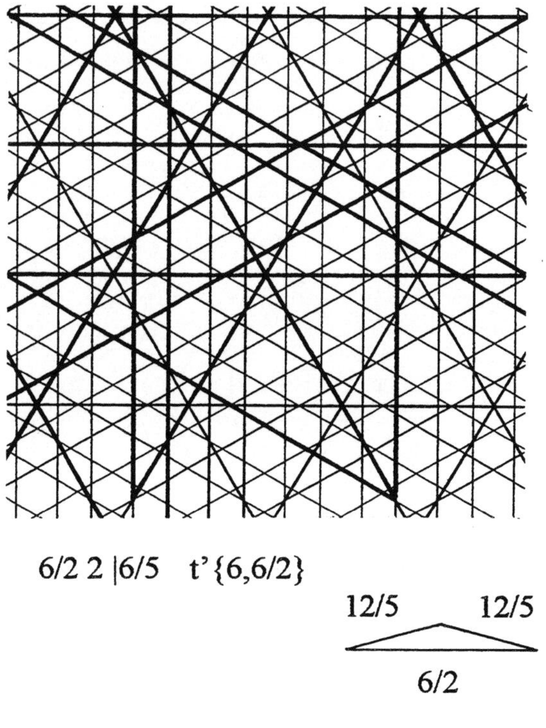

Fig.52: Quasitruncated Star Tilings

EVEN FACED STAR TILINGS

Having explored the first truncation series for this new family of star tilings, located along the two short sides of the Schwarz triangle, we must now turn to the centre of the triangle and look for the truncation of the quasi-regular star tiling.

This will be a 6/2 2 6|, and will accordingly have at each vertex a square, a dodecagon and a double hexagon. It is shown in Figure 53 and actually looks like a 2| 3 6 rhombi tiling, but has no triangular faces, only triangular holes.

It is in fact, as might be expected by now, three overlaid 3 2 6| even faced tilings, each double hexagon centred within a dodecagon, each square covered by two dodecagons and each triangular hole covered by three dodecagons making the density up to three everywhere.

QUASI EVEN FACED STAR TILINGS

Taking this process to its logical conclusion, there are of course also two quasitruncated and one quasiquasitruncated even faced star tilings in this same family.

The latter is shown below in Figure 54 and, although it may not be clear, is actually three overlaid quasiquasitruncated tilings (3/2 2 6/5|) from the original family .

The two slightly simpler quasitruncated even faced star tilings are shown opposite in Figure 55:

6/2 2 6/5| is again more complicated whilst 6/4 2 6| has double {6/2} faces (density –4) concentric with one dodecagon and overlapped by a further six dodecagons (density +7), again making a total of three.

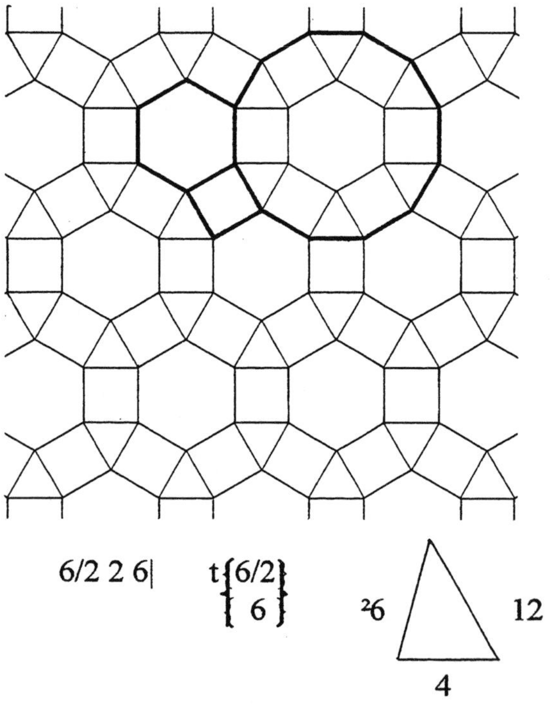

Fig.53: An Even Faced Star Tiling

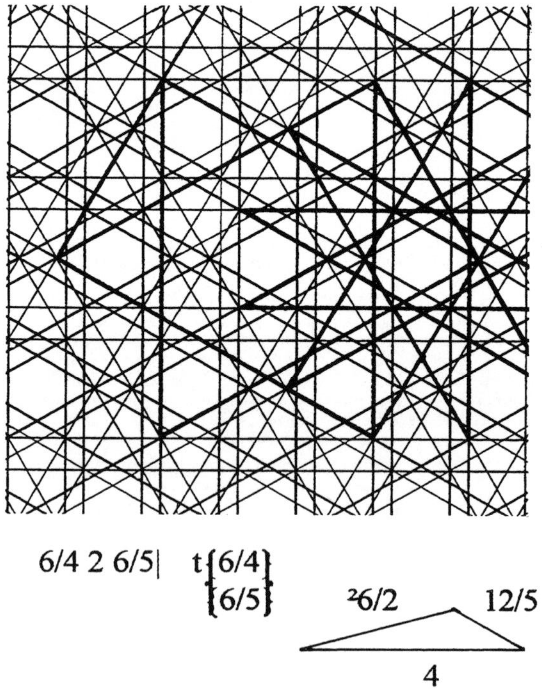

Fig.54: A Quasiquasitruncated Star Tiling

RHOMBI STAR TILINGS

The even faced star tilings just described are first 'truncations' in a second series which, as before, lead on to second 'truncations', the rhombi and quasirhombi star tilings.

The normal and quasiquasitruncated versions lead to the 6/2 6 |2 rhombi star tiling shown below in Figure 56, consisting of three overlaid 3 6 |2 rhombi tilings. This looks like the even faced 6/4 2 6| but has no dodecagon faces, much in the same way that the 3 6 |2 rhombi tiling had a lookalike quasi even faced tiling 3/2 2 6|.

The two quasitruncated even faced star tilings lead to the 6/4 6 |2 quasirhombi star tiling shown overleaf in Figure 57. This in its turn looks like a 6/4 2 6/5|.

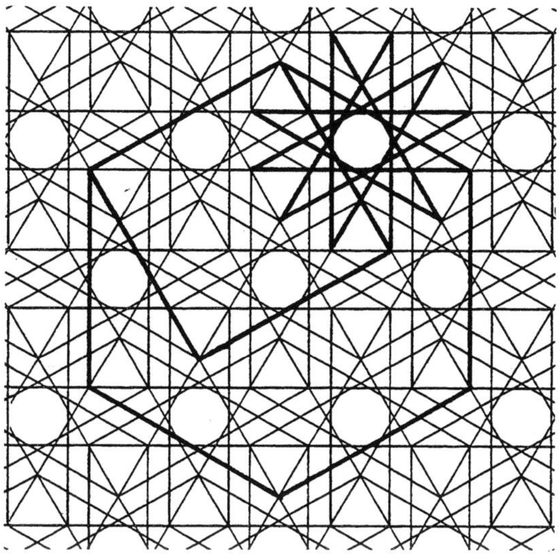

6/2 2 6/5| t{6/2, 6/5}

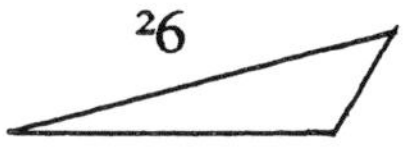

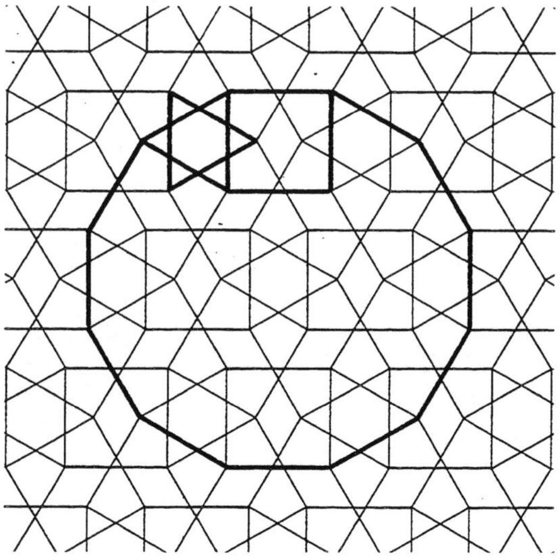

6/4 2 6| t{6/4, 6}

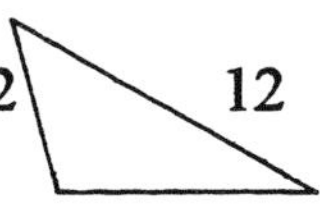

Fig.55: Quasi Even Faced Star Tilings

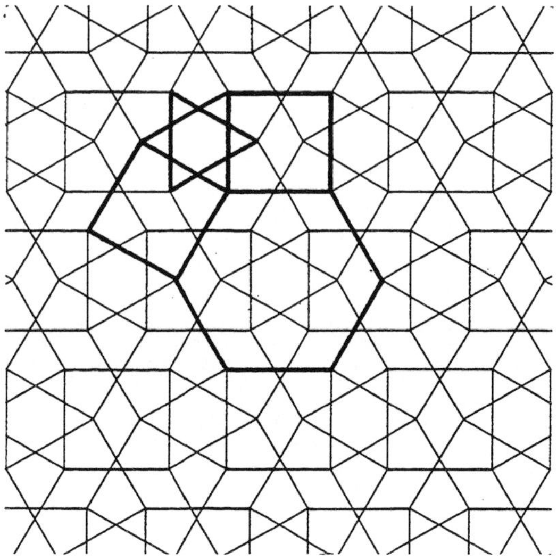

6/2 6 |2 r{6/2, 6}

Fig.56: A Rhombi Star Tiling

SNUB STAR TILINGS

By now it should come as no surprise that there are in fact also both a snub and a quasisnub star tiling formed by twisting the square faces of the rhombi and quasirhombi star tilings respectively into two triangles.

These are shown in Figure 58 and again have a density of three, consisting of three overlaid equivalents from the (3 2 6) family.

The members of the (6/2 2 6) family have thus consistently comprised three overlaid equivalent tilings from the (3 2 6) family, each layer connected to the others by the {6/2} faces (or their truncations $\{^2 6\}$). These are effectively compound tilings resulting from the inclusion of compound polygons at the outset. They form nevertheless a complete? family, as summarised in Figure 59, and have given rise to many further interesting patterns in the plane.

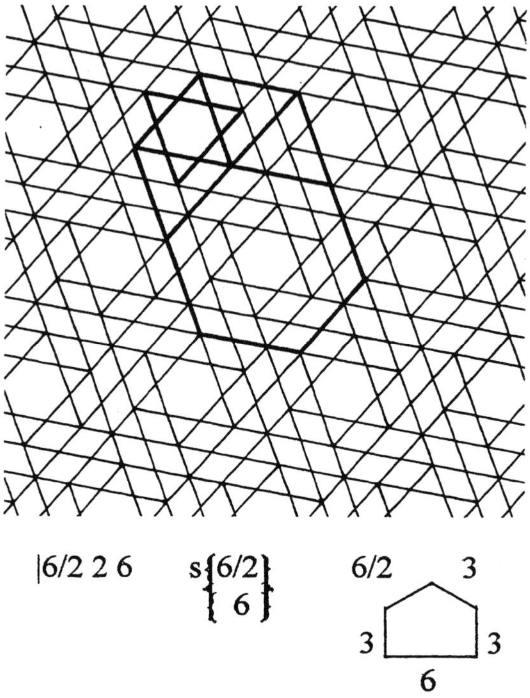

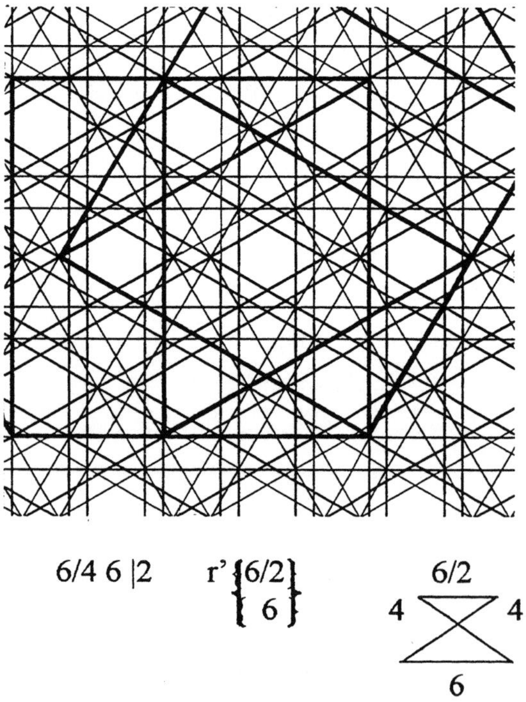

Fig.57: A Quasirhombi Star Tiling

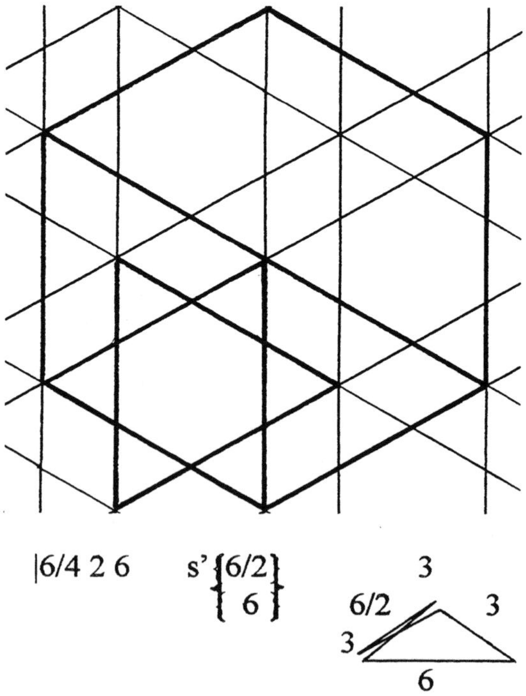

Fig.58: Snub and Quasisnub Star Tilings

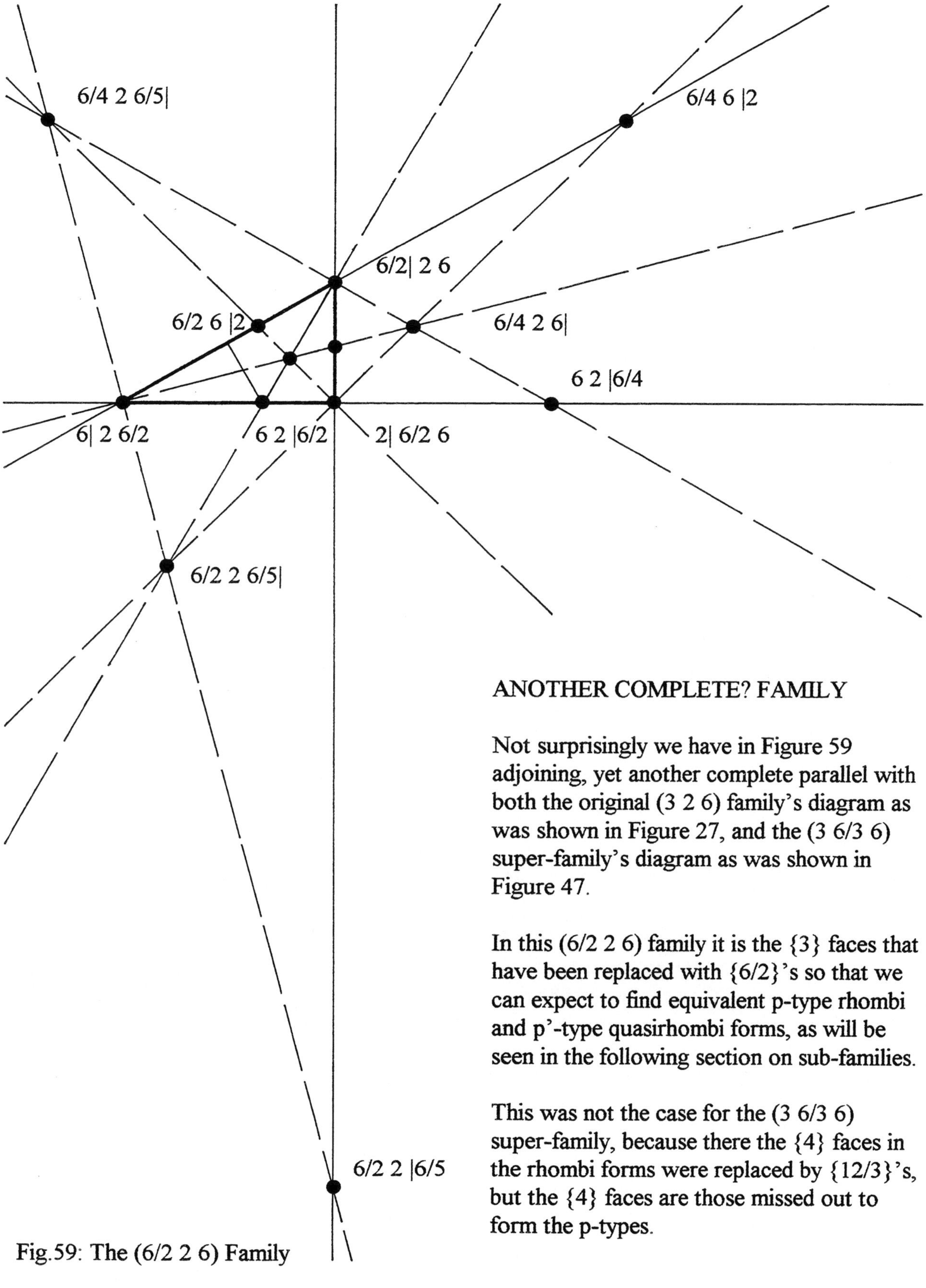

Fig.59: The (6/2 2 6) Family

ANOTHER COMPLETE? FAMILY

Not surprisingly we have in Figure 59 adjoining, yet another complete parallel with both the original (3 2 6) family's diagram as was shown in Figure 27, and the (3 6/3 6) super-family's diagram as was shown in Figure 47.

In this (6/2 2 6) family it is the {3} faces that have been replaced with {6/2}'s so that we can expect to find equivalent p-type rhombi and p'-type quasirhombi forms, as will be seen in the following section on sub-families.

This was not the case for the (3 6/3 6) super-family, because there the {4} faces in the rhombi forms were replaced by {12/3}'s, but the {4} faces are those missed out to form the p-types.

TWO MORE SUB-FAMILIES

In the same way that the (3 2 6) Möbius triangle was doubled up to produce two new Schwarz triangle sub-families, the analogous (6/2 2 6) Schwarz triangle can be doubled up to produce two further sub-families.

These are (6/2 6/2 6/2) and (6 6/4 6) as shown in Figure 60. These Schwarz triangles each consist of six Möbius triangles and can be expected to produce tilings with a density of six.

As before the tilings predicted on the central axes of the triangles will simply be doubled up versions of the tilings in the parent family. These themselves were already three dense, hence we achieve the density of six predicted.

The tilings concerned are for the first sub-family: 6/2| 6/2 6/2 at the apex,
6/2 6/2 6/2| at the centre,
6/2 6/2| 6/2 on the base,
and finally 6/4 6/2 6/4| in the centre below;

whilst for the second sub-family they are:
6/4| 6 6 at the apex
6 6/4 6| at the centre
6 6 |6/4 on the base
and finally 6/5 6/4 6/5| in the centre below.

It is worth noting here that the (6/2 6/2 6/2) sub-family will produce little other than these on account of the symmetry of its Schwarz triangle. The only other unspecified member of this sub-family is 6/2 6/4 |6/4, which is a coincident tiling.

The next couple of pages will thus be devoted to the more productive (6 6/4 6) sub-family, and both sub-families will be summarised in Figures 65 and 66 following on page 61.

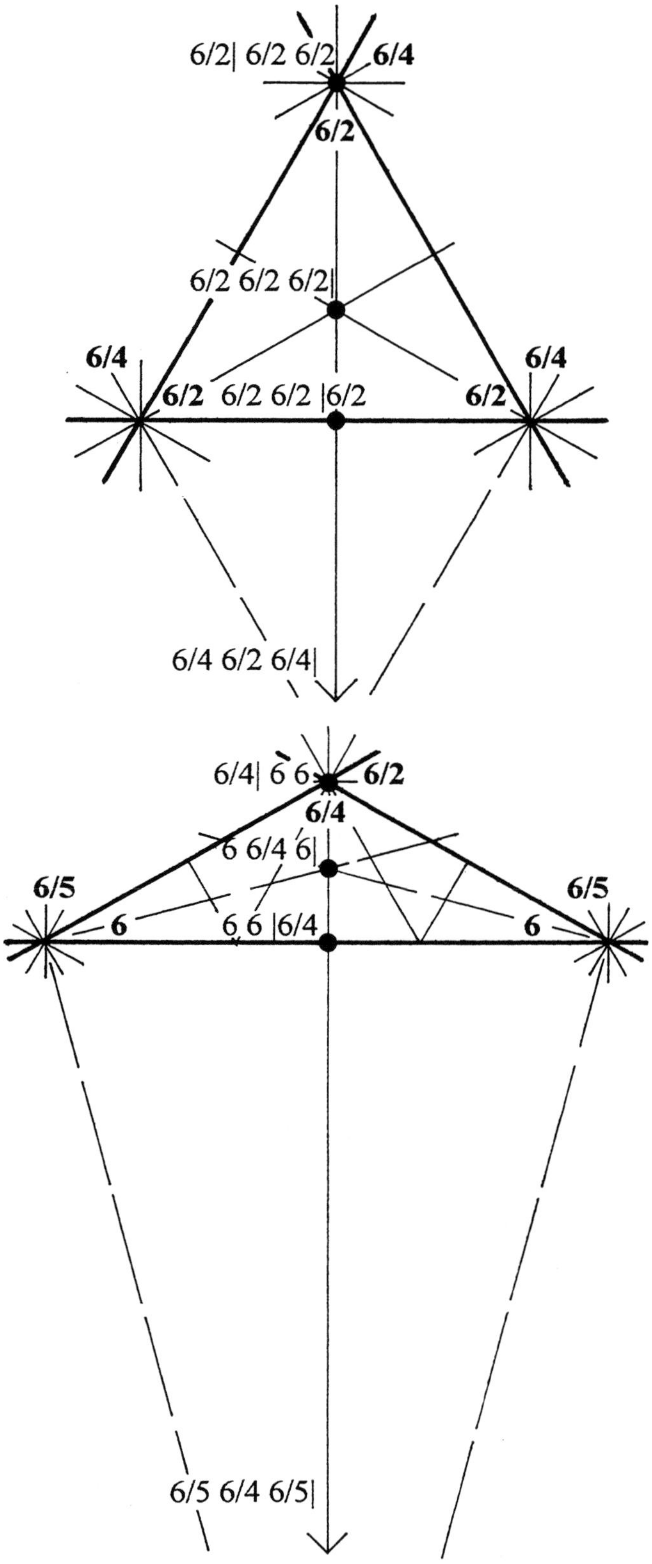

Fig.60: Sub-Family Schwarz Triangles

A QUASI-REGULAR TILING

The lower vertices of the (6/2 6/2 6/2) sub-family produce 6/2| 6/2 6/2, which is the same as that from the apex, a double tiling.

However the lower vertices of (6 6/4 6) produce the more interesting tiling, 6| 6/4 6, which is shown below in Figure 61.

This quasi-regular tiling has a vertex figure that is theoretically dodecagonal with {6} and {6/4} faces alternating around it. The latter inverse faces can be twisted out to produce a dodecagram with alternating {6} and {6/2} faces as shown below.

The resultant tiling actually consists of three overlaid 6| 3/2 6 tilings, which themselves are already double dense, so that the final density totals the expected six.

A RHOMBI TILING

Similarly the upper edges of (6/2 6/2 6/2) only produce 6/2 6/2 |6/2, which has already been considered, whilst in the other sub-family we find 6 6/4 |6, which is again more interesting.

This tiling consists of three overlaid 6 3/2 |6 tilings and looks the same as 6/2 6 |2, the rhombi tiling from the {6/2,6} family. It is shown below in Figure 62 along with its vertex figure, which again includes a {6/4} face twisted into its inverse {6/2}.

In reality this tiling is actually the p-type rhombi tiling from the {6/2,6} family and has no square faces. The centre of each apparent square is covered by four dodecagons and two hexagons, maintaining the density at six.

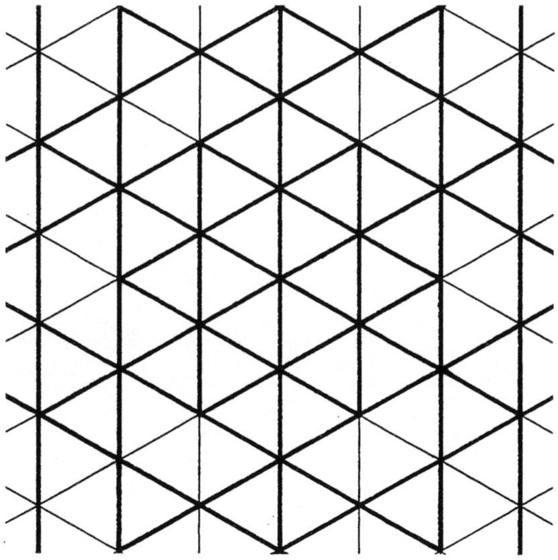

6| 6/4 6

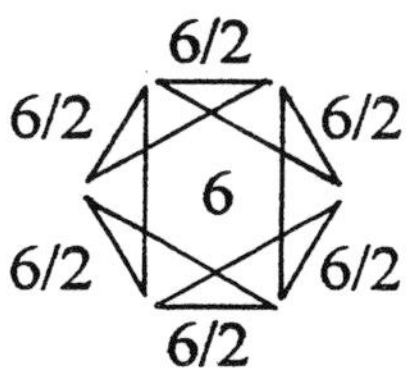

Fig.61: A Quasi-Regular Tiling

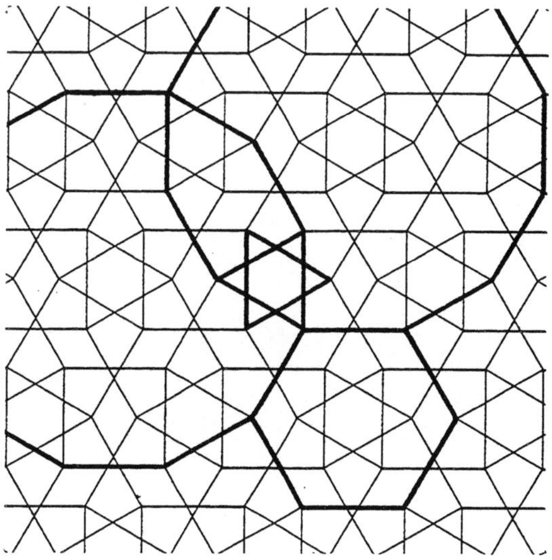

6 6/4 |6 $p\begin{Bmatrix}6/2\\6\end{Bmatrix}$

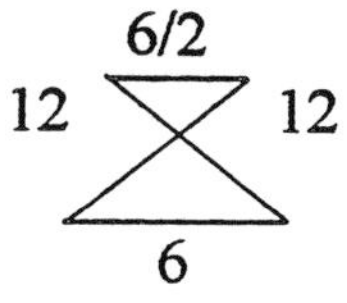

Fig.62: A Rhombi Tiling

AN EVEN FACED TILING

Moving off the Schwarz triangles into the region at top right or left we can find the quasitruncated even faced tilings 6/2 6/4 6/4| and 6 6/2 6/5|. The first of these has again already been considered, whilst the second once more warrants further investigation.

This latter is shown below in Figure 63 and predictably will have a double hexagon, a dodecagram and a dodecagon at each vertex.

It consists of three overlaid 6 3 6/5| tilings. Each dodecagram (density –5) is concentric with a double hexagon (–2) and a dodecagon (+1), and is overlapped by a further twelve dodecagons, bringing the total density to the expected six.

A QUASIRHOMBI TILING

Finally moving to the extended upper edge of our Schwarz triangles we find for the first sub-family the already mentioned coincident tiling 6/2 6/4 |6/4 (as with all coincident tilings the first two numbers in the symbol are inverses).

For the other family we find a proper quasirhombi tiling, 6 6/2 |6/5, which is shown below in Figure 64. In parallel to its rhombi partner it consists of three overlaid 6 3 |6/5 tilings and is actually the p'-type quasirhombi tiling for the {6/2,6} family.

Both this tiling and its rhombi counterpart have no square faces so that, as we found before, the sub-family will not produce any conventional snubs.

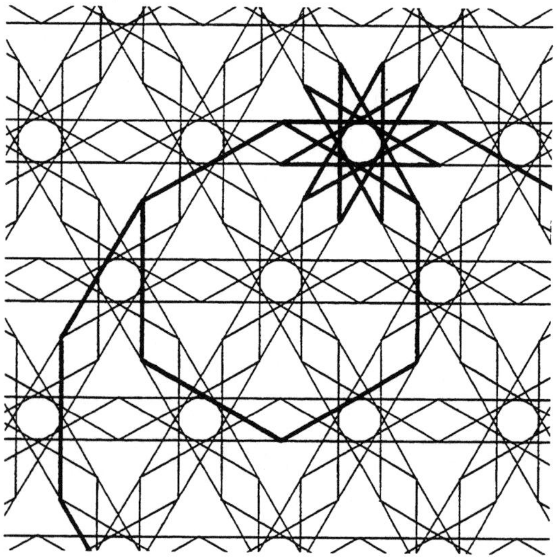

6 6/2 6/5|

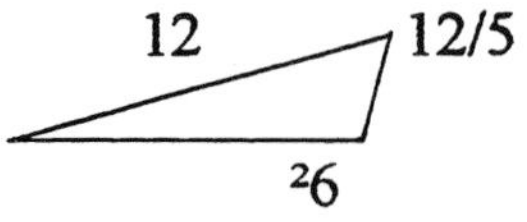

Fig.63: An Even Faced Tiling

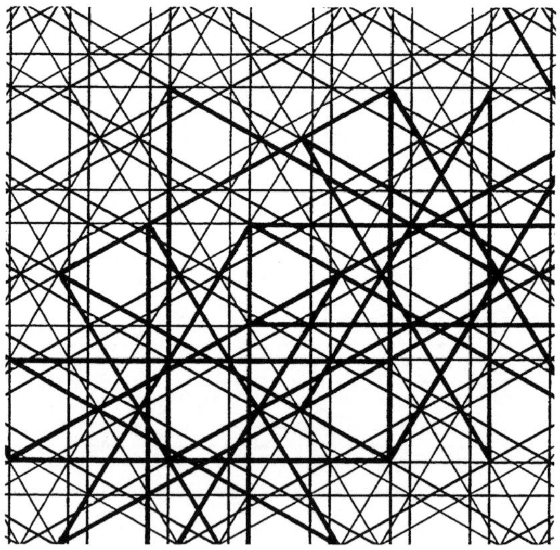

6 6/2 |6/5 p' {6/2 / 6}

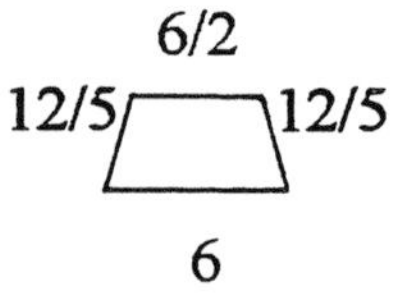

Fig.64: A Quasirhombi Tiling

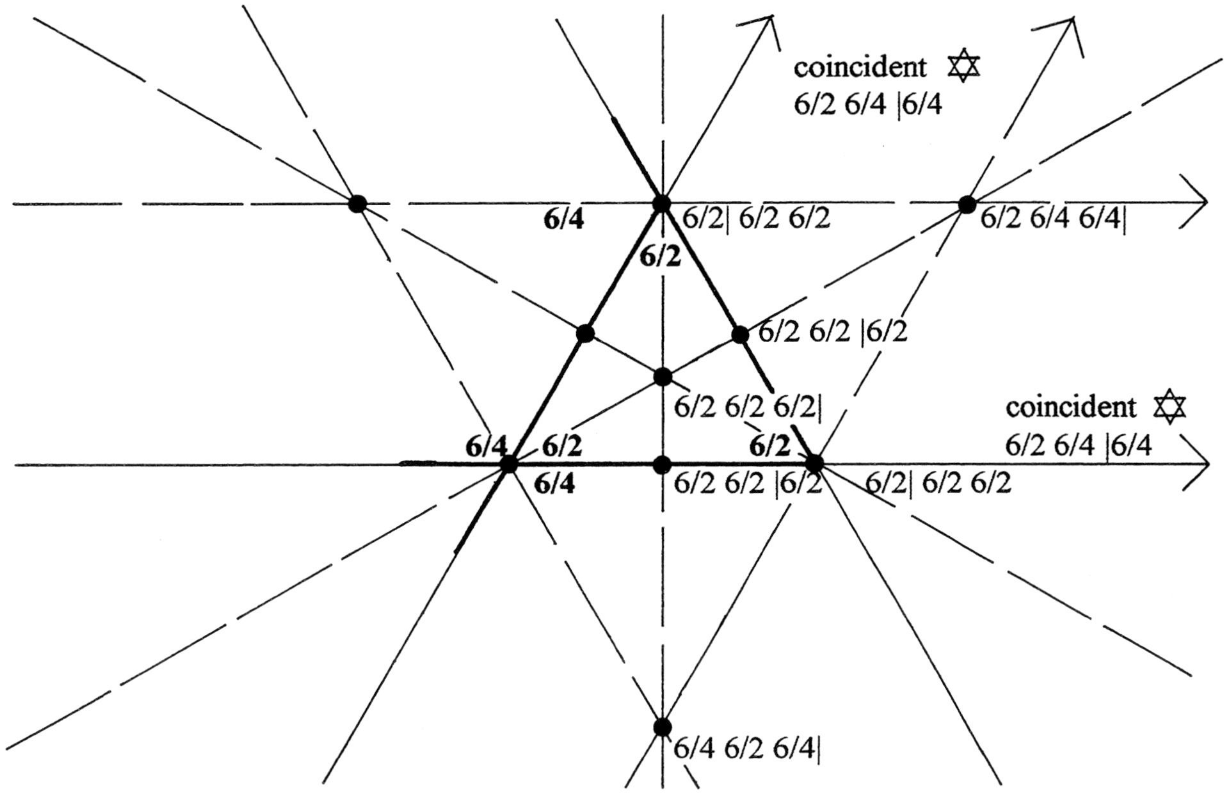

Fig.65: The (6/2 6/2 6/2) Sub-Family

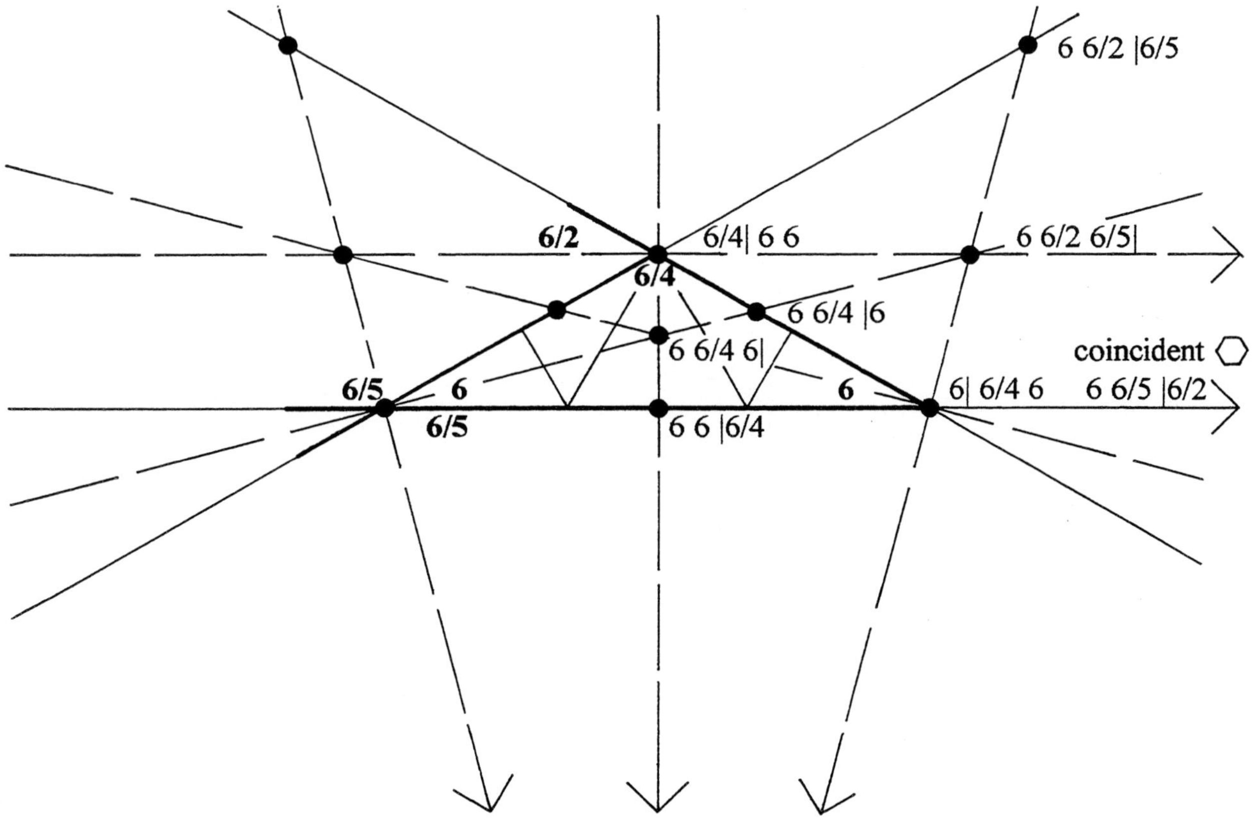

Fig.66: The (6 6/4 6) Sub-Family

6/5 6/4 6/5|

ANOTHER SUPERFAMILY

We have already seen how the (3 2 6) family of tilings became the (3 6/3 6) super-family of density four, when the digon {2} was replaced by the {6/3} cross polygon face.

We have also seen how the (3 2 6) family became the (6/2 2 6) family of star tilings of density three, when the triangle {3} was replaced by the star polygon {6/2}.

We can in fact substitute both compound polygons for the original polygons and end up with a (6/2 6/3 6) Schwarz triangle as shown below in Figure 67.

This Schwarz triangle can also be seen to consist of the two star tiling sub-family triangles, (6/2 6/2 6/2) and (6 6/4 6), placed back to back to form a new super-family.

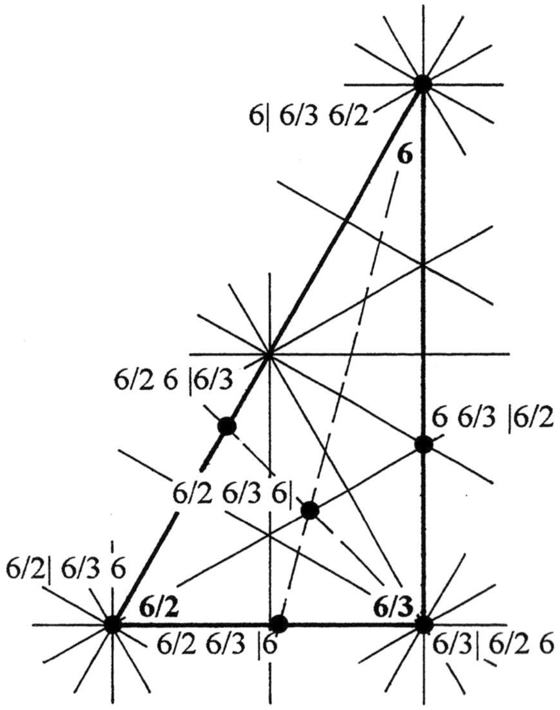

Fig.67: (6/2 6/3 6) Schwarz Triangle

THREE QUASI-REGULAR TILINGS

The members of this new super-family of tilings have density twelve, and can be seen as either threefold versions of the (3 6/3 6) super-family or fourfold versions of the (6/2 2 6) family. They can thus be regarded as doubly compound.

The vertices of the new Schwarz triangle produce three further quasi-regular tilings as shown opposite in Figure 68.

6/3| 6/2 6 appears at first to be a simple triangular tiling, however each apparent triangle is actually only the apex triangle of a {6/2} face (density 1), as well as a central triangle in another {6/2} (density 2) and is overlaid by nine hexagons, three along each of three symmetrical directions. It is as predicted a threefold 6/3| 3 6 and a fourfold 2| 6/2 6, which overall makes it a twelvefold 2| 3 6. Like 6/3| 3 6, it has a three rectangle vertex figure, which is normalised into a {12/3} when truncated.

6/2| 6/3 6 also appears to be a triangular tiling, but is in fact twelve overlaid 3| 2 6 hexagon tilings, any one apparent triangle being part of twelve different hexagons. This tiling can also be seen as three 3| 6/3 6 tilings or four 6/2| 2 6 tilings. This tiling has an unusual vertex figure, which normalises into a double hexagon on truncation.

6| 6/3 6/2 appears more complex, but consists of twelve overlaid 6| 2 3 triangle tilings, which once again can be broken down into three 6| 6/3 3 tilings or four 6| 2 6/2 tilings.

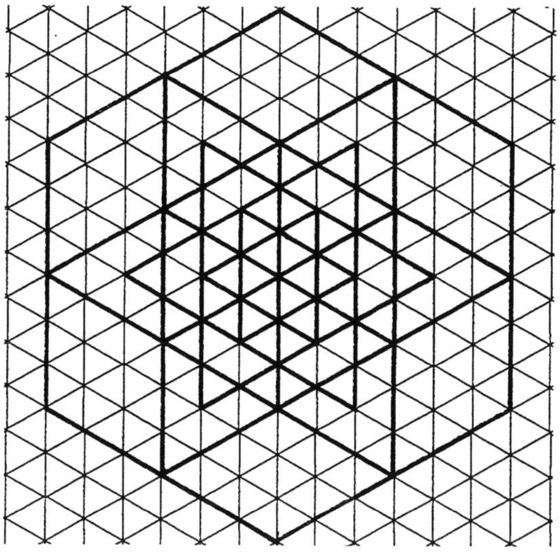

6/3| 6/2 6

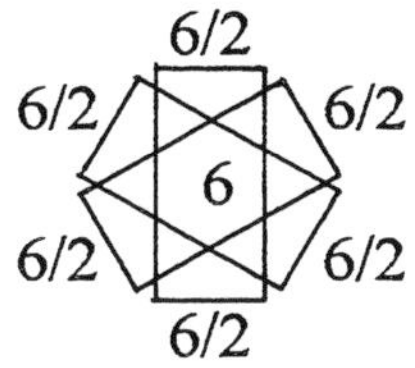

6/2| 6/3 6

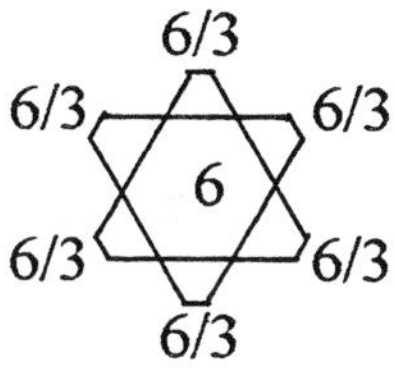

6| 6/3 6/2

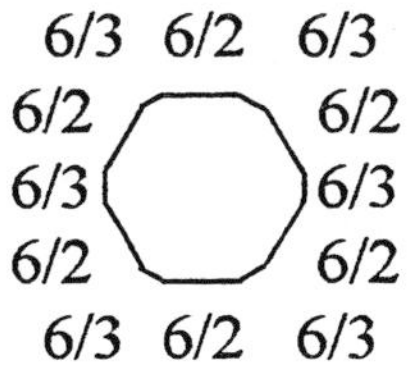

Fig.68: Three Quasi-Regular Tilings

FOUR EVEN FACED TILINGS

Moving on to the centre of the new Schwarz triangle we can find the 6/2 6/3 6| even faced tiling, which will have a triangular vertex figure giving a $\{^2 6\}$, {12/3} and {12} face at each vertex. This is shown overleaf in Figure 69 and looks the same as 3 6 |6/3, but has no triangular faces. Each triangular hole is here covered by 6 dodecagons (density 1) and 3 double hexagons (density 2), making our obligatory total of density twelve.

Also shown in Figure 69 are the three quasi even faced tilings from the outlying positions on our Schwarz triangle. These are the quasi even faced tilings 6/4 6/3 6| and 6/2 6/3 6/5| along with the double quasitruncated version 6/4 6/3 6/5|. All four consist of twelve overlaid equivalent tilings from the (3 2 6) family.

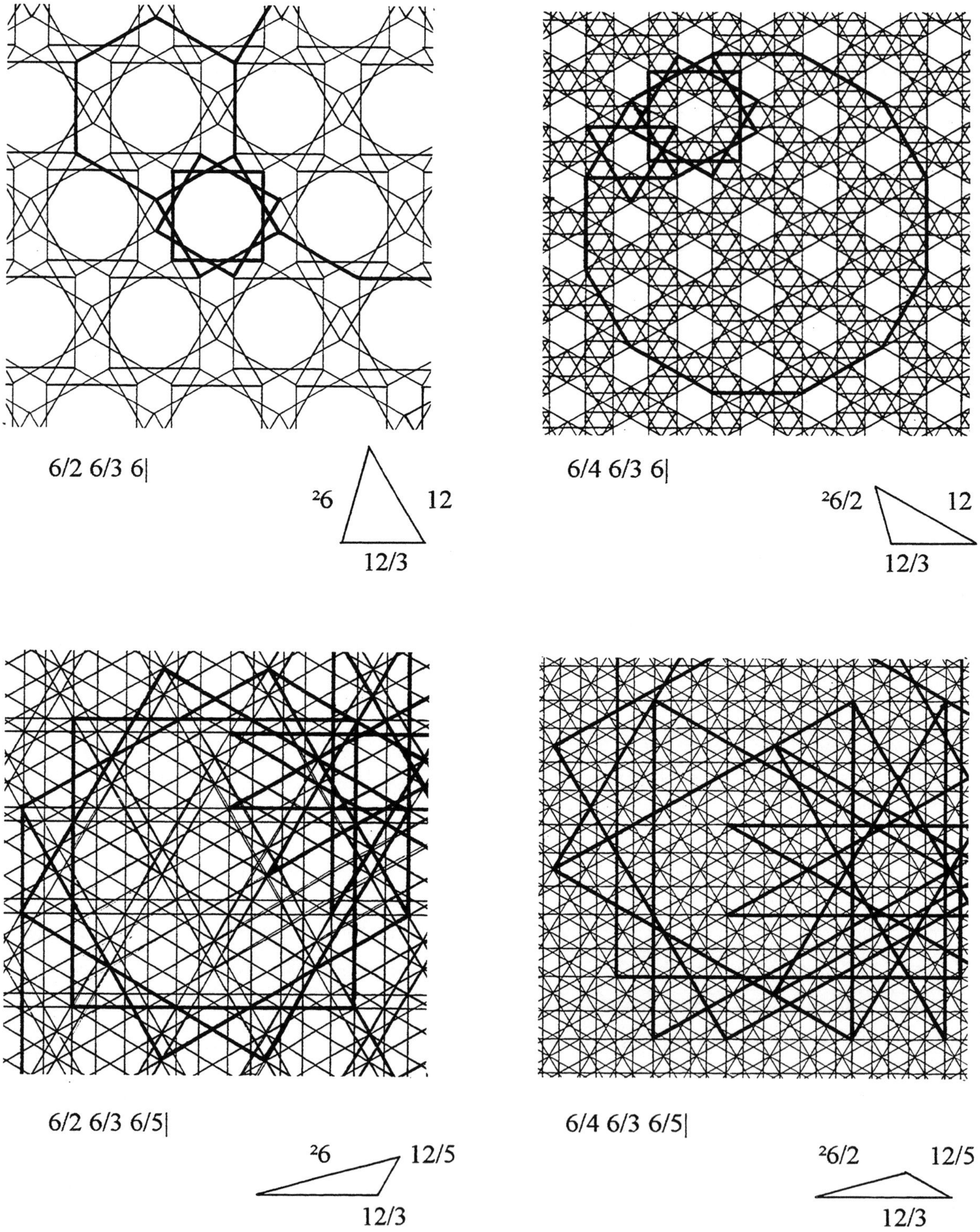

Fig.69: Four Even Faced Tilings

SIX RHOMBI TILINGS

As with our previous super-family we now come to examine the products of the edges of our new Schwarz triangle.

Again on the hypotenuse we find the true 'rhombi' form 6/2 6 |6/3, which is essentially a 6/2 6 |2 with its square faces replaced by {12/3} faces. It thus consists of either four of the latter tilings overlaid or alternatively three 3 6 |6/3 tilings overlaid, or even twelve 3 6 |2 tilings. It is shown in Figure 70 where we can see that any {6/2} face is concentric with both a {12/3} and a {6}, and is overlapped by a further six hexagons, making up the total density to the expected twelve.

On the extended edge of the hypotenuse we find our equivalent 'quasirhombi' form 6/4 6 |6/3 which is also shown in Figure 70 and consists of twelve quasirhombi 3/2 6 |2 tilings overlaid. As we found with the earlier families, both of these forms are lookalikes sharing a basic grid pattern with one of the even faced tilings (see opposite).

This super-family, like the (3 6/3 6) super-family, does not produce p-type rhombi or p'-type quasirhombi tilings other than those already seen for the (6/2 2 6) family. Also, as before, if we try to form sub-families from the (6/2 6/3 6) Schwarz triangle, we can only get larger versions of the previously seen (6/2 6/2 6/2) or (6 6/4 6) sub-families of the (6/2 2 6) family.

Overleaf in Figure 71 the products of the other two edges of our Schwarz triangle are shown, along with their quasitruncated equivalents. These are again, as might by now be expected, twelvefold versions of the original regular tilings' truncations and quasitruncations.

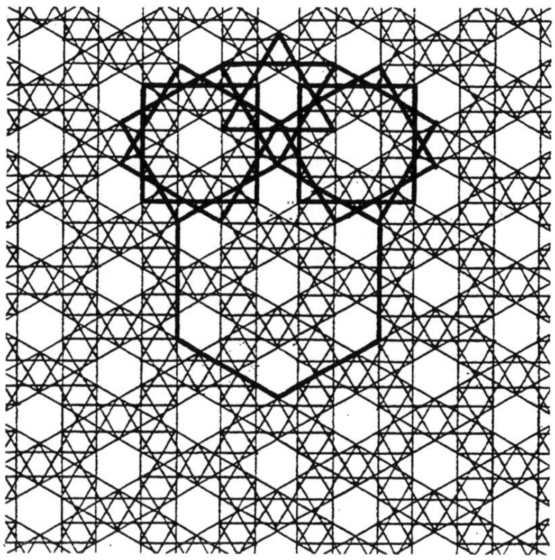

6/2 6 |6/3

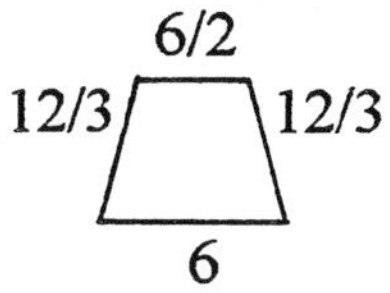

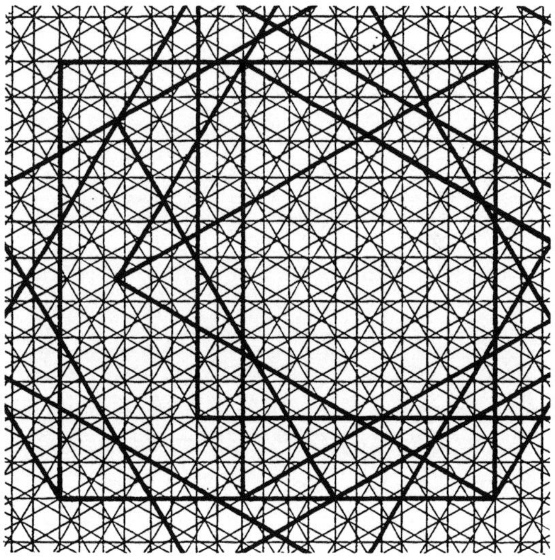

6/4 6 |6/3

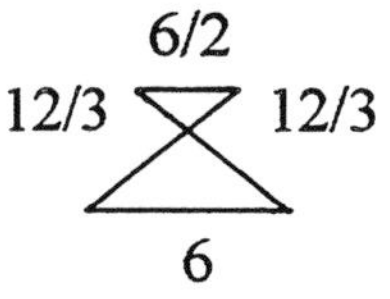

Fig.70: Rhombi and Quasirhombi Tilings

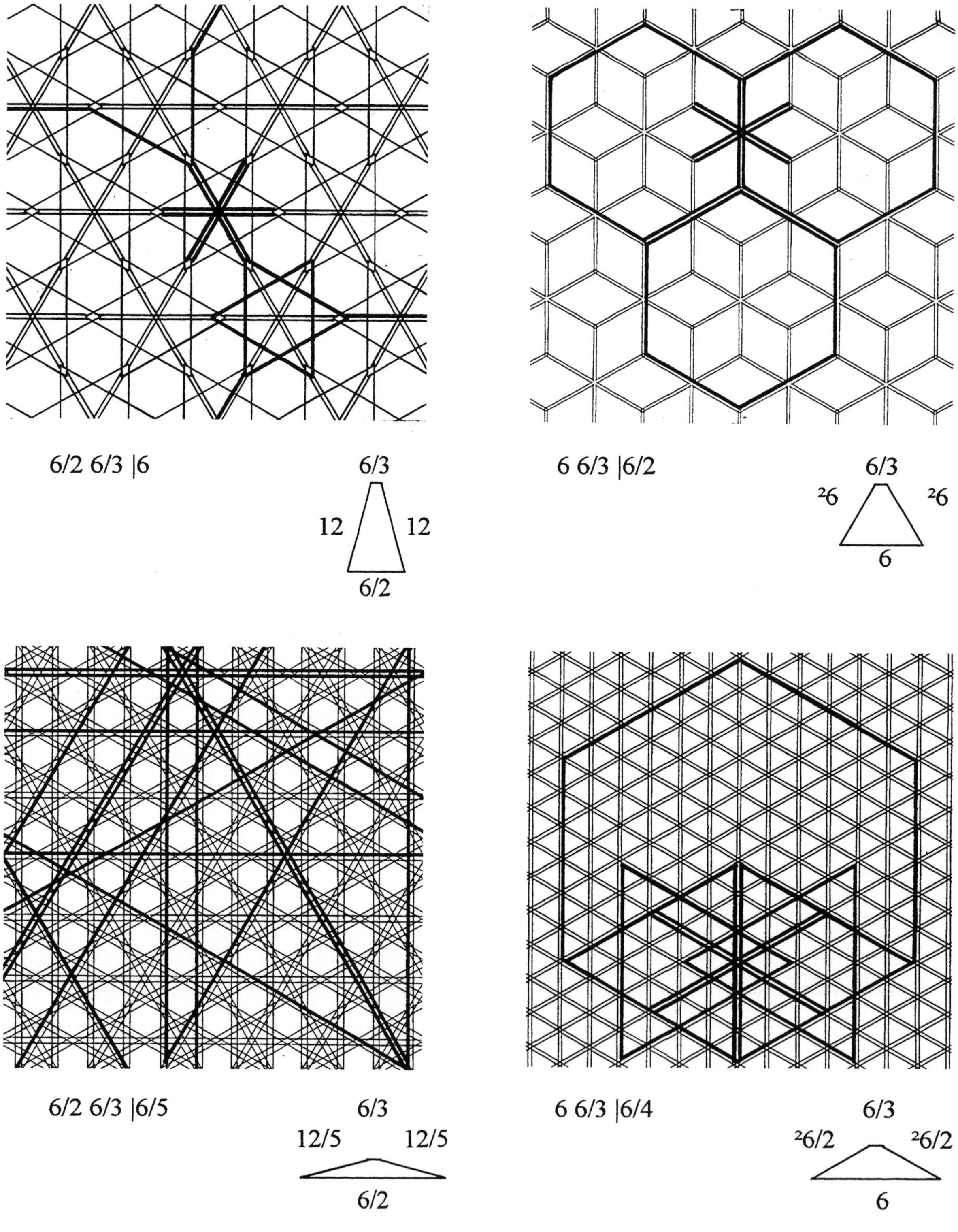

Fig.71: Truncated & Quasitruncated Tilings

TWO SNUB TILINGS

The snub tilings of our previous super-family (3 6/3 6) were fourfold versions of the snub tilings of the original (3 2 6) family. In a similar way the snub tilings of the present super-family (6/2 6/3 6) are fourfold versions of the snub tilings in the (6/2 2 6) star tiling family, themselves threefold versions of the originals.

The resultant snub and quasisnub tilings are shown in Figure 72 and can also be regarded as overlays of twelve of the appropriate rhombi or quasirhombi tilings, as shown in Figure 70, to each of which the snub 'twist' has been added.

These snubs thus complete this super-family with a full complement of tilings, as shown overleaf in Figure 73. We have now reached the maximum complexity the tilings have to offer, as further combinations of Möbius triangles lead only to repeats of the Schwarz triangles already examined, with but one exception.

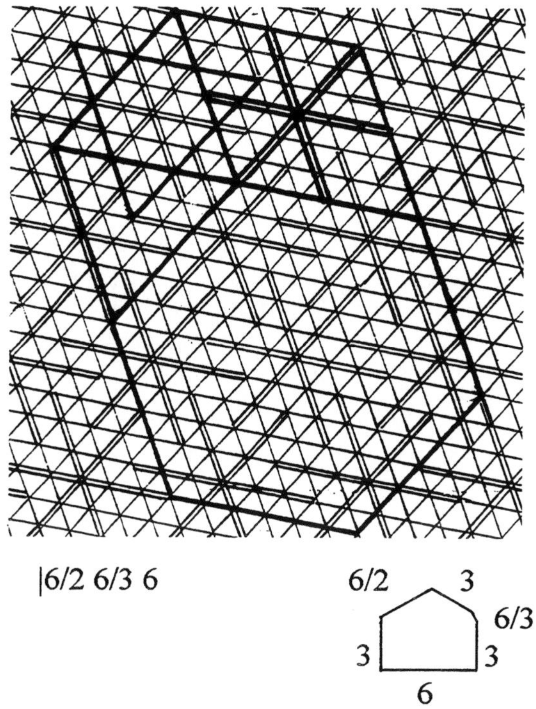

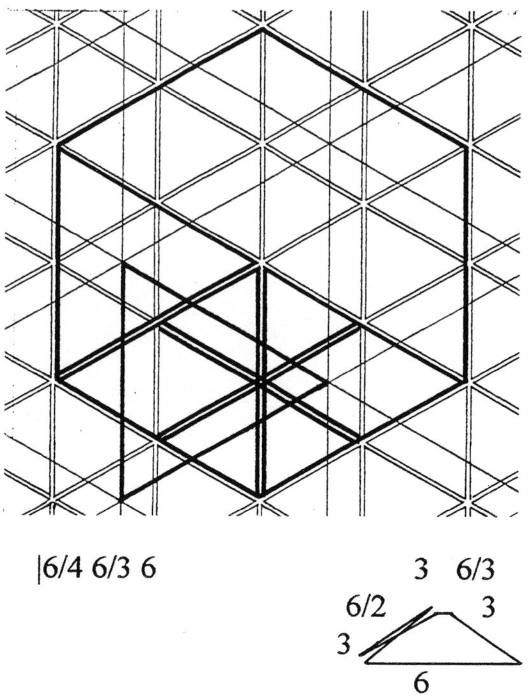

Fig. 72: Snub and Quasisnub Tilings

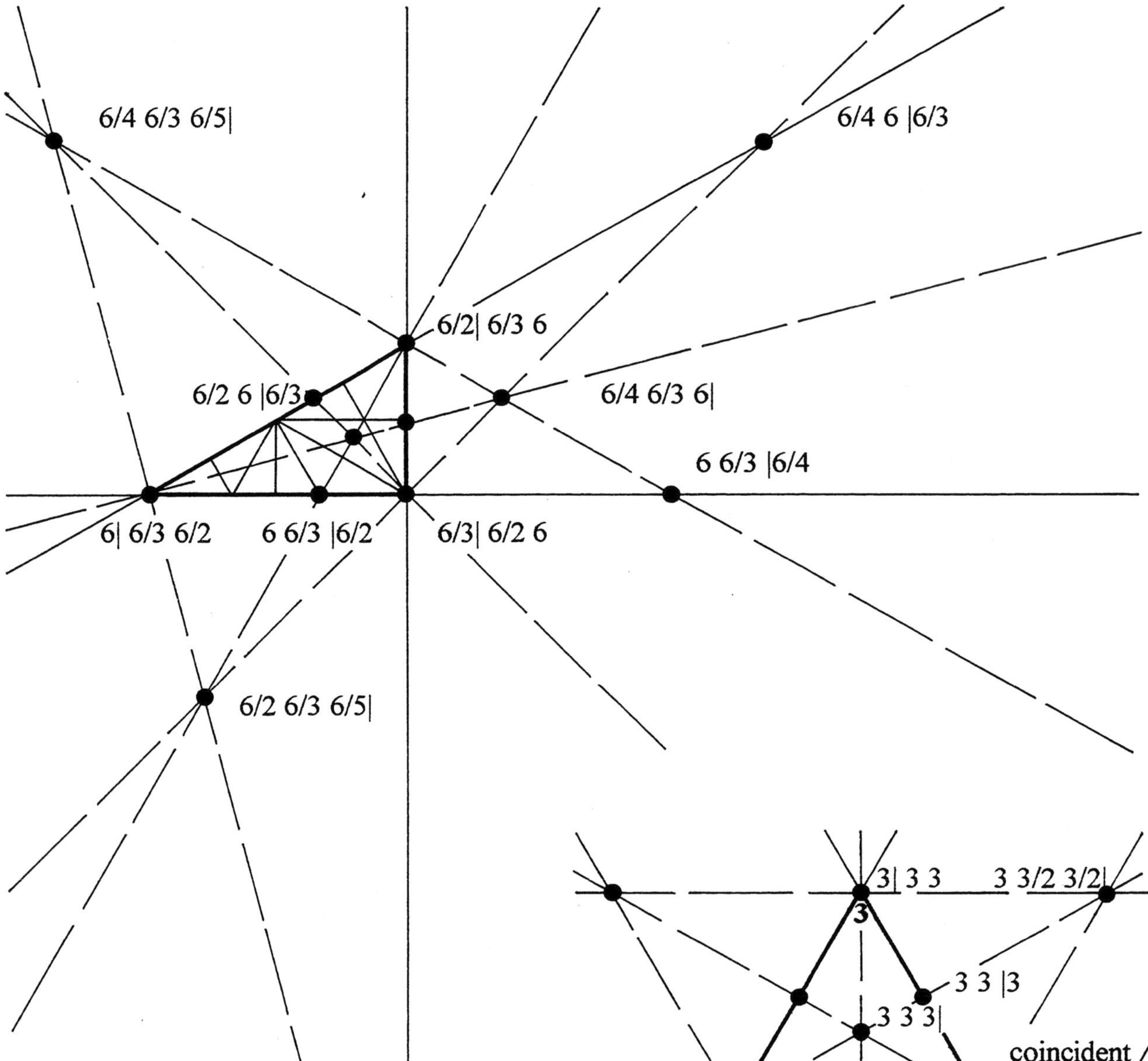

Fig.73: The (6/2 6/3 6) Super-Family

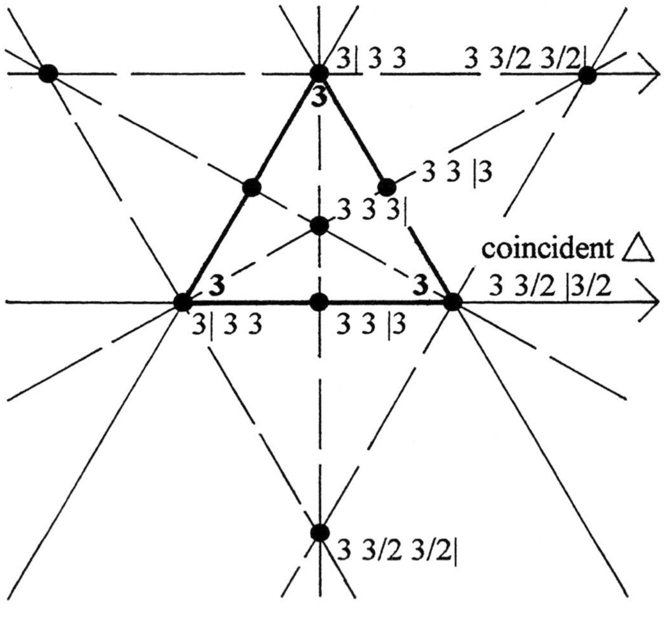

Fig.74: The (3 3 3) Family

THE (3 3 3) FAMILY

There remains one arrangement of Möbius triangles within our original grid that has not yet been considered. It is composed of eighteen (3 2 6) triangles to make up the (3 3 3) Schwarz triangle and is shown in Figure 75.

Essentially we are back to square one, as this does not benefit from consideration as an eighteen fold version of anything, but rather can be considered as a basic Möbius triangle in its own right, part of a simple triangular grid also shown in Figure 75.

Considered this way we find: that 3| 3 3, at the vertex, is itself a triangular tiling equivalent to 6| 2 3; that 3 3 |3, on the edge, is a hexagon-triangle tiling equivalent to 2| 3 6; and that 3 3 3|, at the centre of the Schwarz triangle, is a hexagon tiling equivalent to 3| 2 6.

The 'quasi' forms are no more interesting: 3/2 3/2 3| is again a triangle tiling but with hidden hexagons equivalent to 2 6 |3/2; whilst 3 3/2 |3/2 is the elusive coincident triangular tiling which we might have expected to encounter sooner. As with some of our earlier sub-families, there are no squares involved in the rhombi and quasirhombi forms so we will not find the usual snub forms either.

The symmetry of this Schwarz triangle means that there are no other possibilities here and we seem to have truly finished our journey through the various possibilities for symmetry families. However, a few loose ends remain to be tied up to make quite certain this is the case.

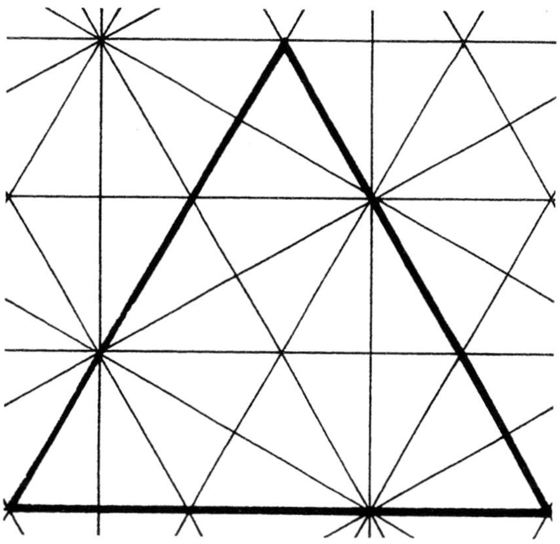

(3 3 3) Schwarz Triangle

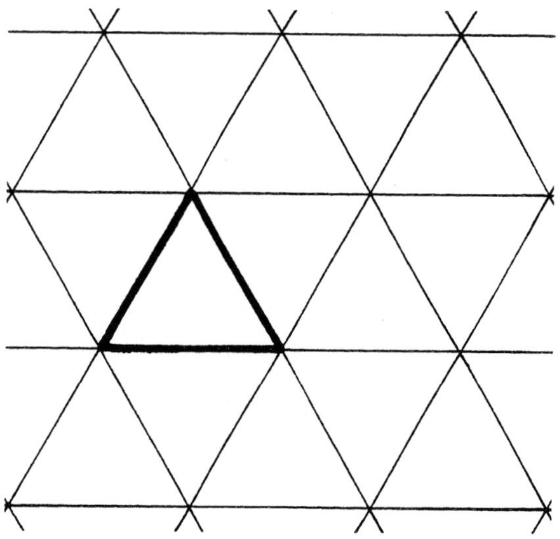

(3 3 3) Möbius Triangle

Fig.75: (3 3 3) Generating Triangles

HEXISNUB TILINGS

In the course of our travels so far, we have seen a number of snub tilings, which in general have had the symbol |2 x y and have been formed by twisting the vertices of the family's rhombi or quasirhombi tilings, x y |2 so that the paired square faces (truncated {2}s) each become two triangles. The vertex figures have thus changed from the isosceles trapezoid (x,4,y,4) to an irregular pentagon (x,3,y,3,3).

We have also seen a number of 'rhombi' type tilings with trapezoid vertex figures, but with paired hexagons and the general symbol x y |3, the hexagons being truncated {3}s. It is possible to apply a similar twist to these to produce a hexisnub tiling which will have the symbol |3 x y. Here the paired hexagons are replaced by four triangles, so that our trapezoid vertex figure (x,6,y,6) becomes an irregular hexagon with the general form (x,3,y,3,3,3). These again have left and right handed enantiomorphic forms, and are characterised by the new triangular faces forming blocks of four within the tiling.

Two examples have already been seen as 'rhombi' snubs, |3 2 6 and |3 6/3 6 and are shown in their new derivation in Figure 76. Here these are both derived from trapezoid vertex figures with paired hexagons either side and a hexagon facing a null face, {2} or {6/3}, the other way.

Three more examples are shown opposite in Figure 77, where there are no null faces. These have all been derived from rectangular vertex figures with paired hexagons either side and combinations of {3} or {6/2} for the other two faces. These give rise to a hexisnub tiling for the (3 3 3) family, one for the (3 3 6/2) sub-family and finally one for a notional (3 6/2 6/2) family.

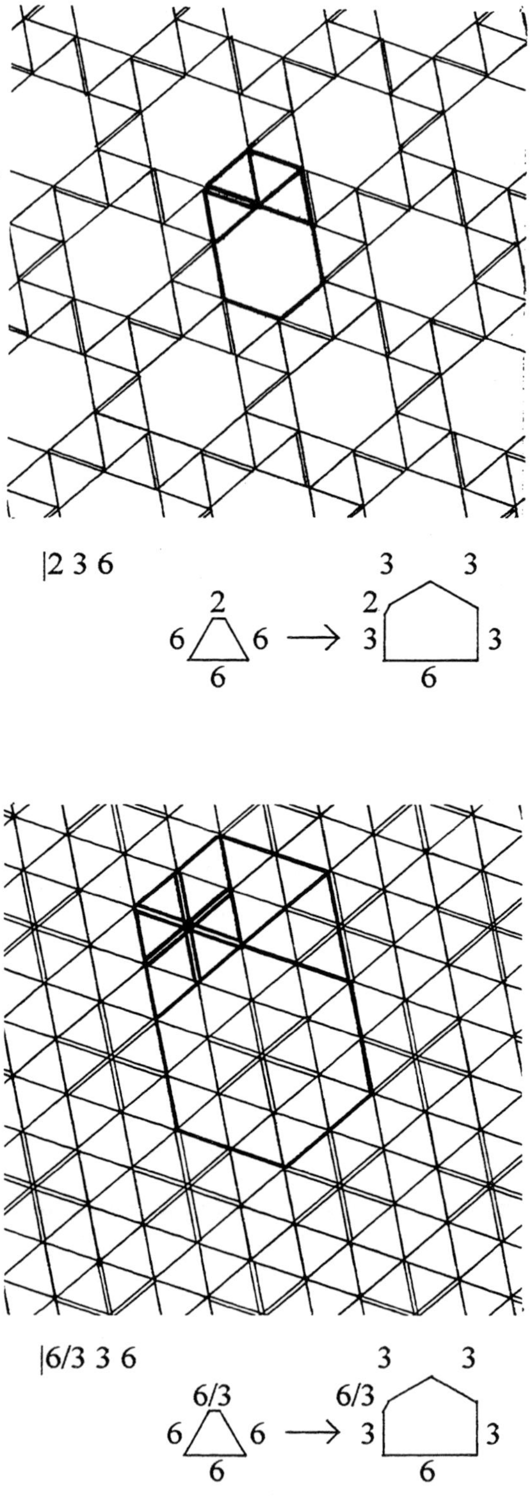

Fig.76: Hexisnub Tilings

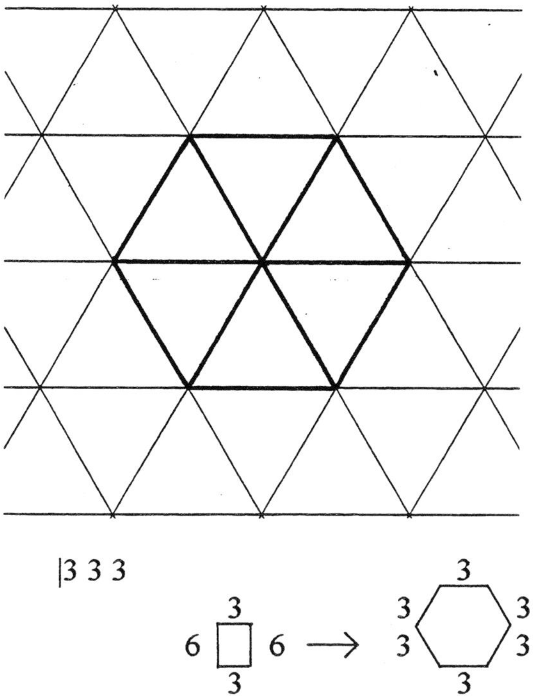

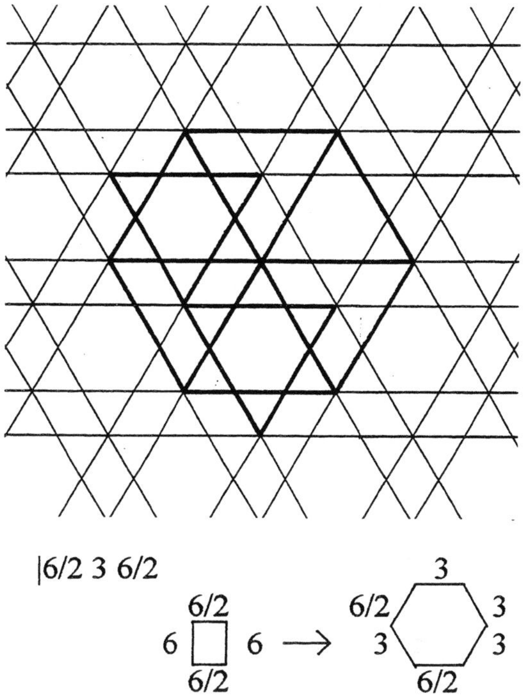

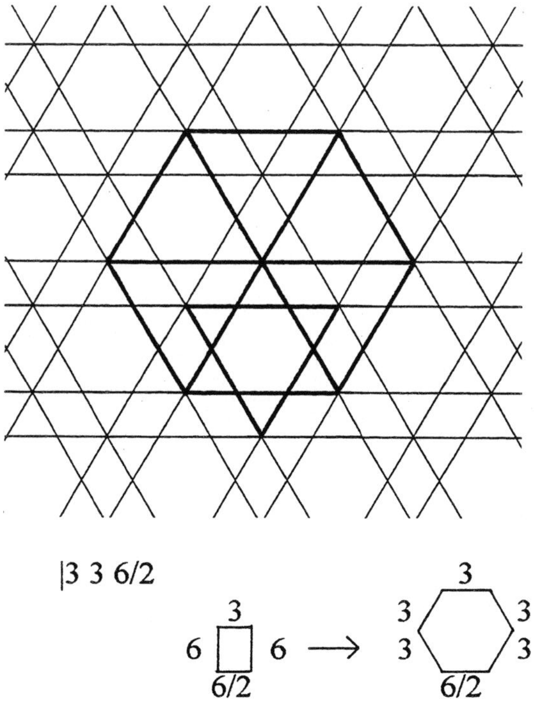

Fig. 77: More Hexisnub Tilings

THE (3 6/2 6/2) FAMILY?

This last hexisnub tiling described was actually derived from the vertex figure of the quasi-regular star tiling, 2| 6/2 6, which if considered as a rhombi type tiling would be given the symbol 6/2 6/2 |3. This is a direct parallel to the 2| 3 6 quasi-regular tiling which we saw on page 69 in another guise as the rhombi 3 3 |3.

There is not however a (3 6/2 6/2) family lurking in the background as might be suspected. It is not possible to construct such a Schwarz triangle within our basic (3 2 6) Möbius triangle grid. This might be met with either disappointment or perhaps just a sigh of relief.

COMPLETING QUADRILATERALS

We saw way back on page 27, that the quadrilateral vertex figures of the rhombi and quasi-rhombi tilings could each be split apart into two triangles, revealing all four of the vertex figures of the even faced tilings. This same operation can be carried out with the corresponding quadrilateral vertex figures of our compound tiling families, but here in one family {6/2} replaces {3}, in the next {12/3} replaces {4}, whilst in the last both are substituted.

We have, however, already seen more triangular vertex figures than just these even faced ones. There were the isosceles triangle vertex figures of the truncations and quasi-truncations of our regular tilings and also the triangular vertex figures of the sub-families' even faced tilings seen on page 39.

Combining these triangles with each other and with those of the original even-faced tilings, we can produce eight further potential 'rhombi' type vertex figures as shown in Figure 78.

The first four of these seem to form rhombi and quasirhombi pairs that suggest the existence of two further tiling families, the (8 2 8/3) and the (12 2 12/5). Although the angles are right to construct a generating Möbius triangle for each of these, it cannot be reflected or rotated so as to fill the plane in a single layer.

The last four shown are even more complicated assemblies of triangles, squares, hexagons and dodecagons, along with their truncations and inverses. These would require even less achievable generating triangles. The general result is that these quadrilateral vertex figures are only actually viable when generated with a double polygon on one of the parallel sides.

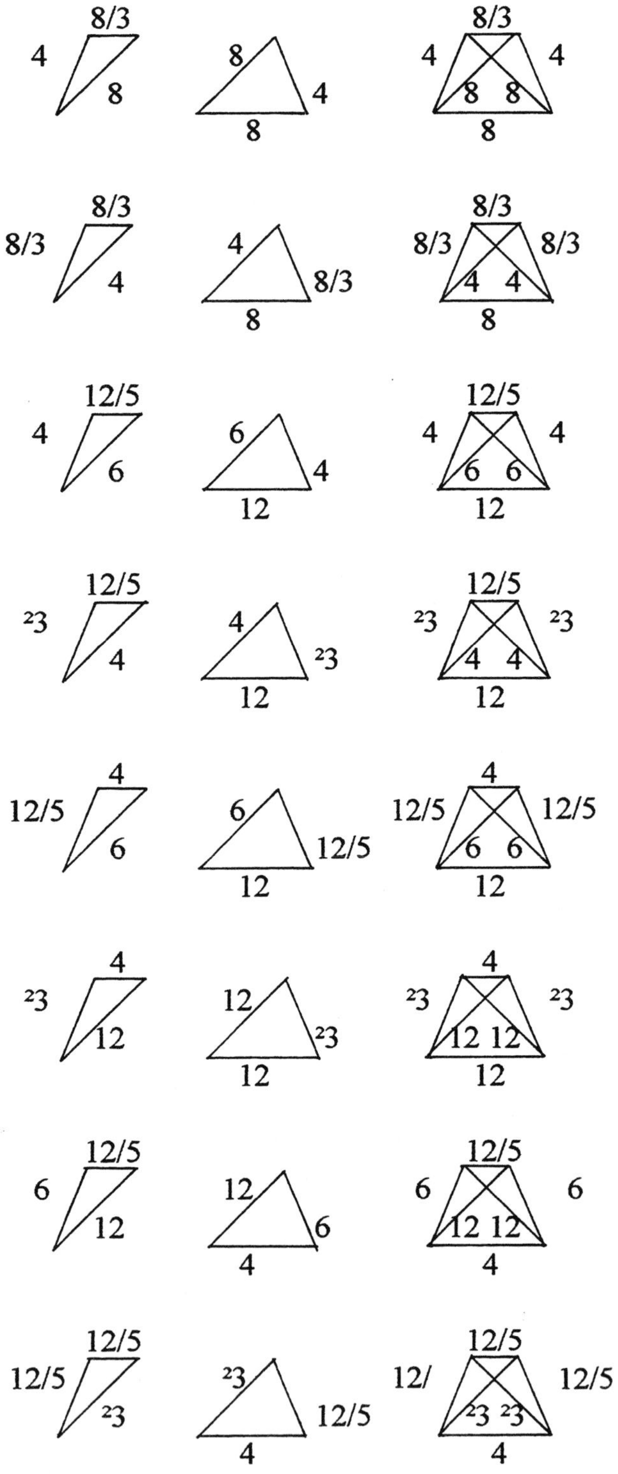

Fig.78: Completing Quadrilaterals

SUMMARY OF TILINGS

We seem to have exhausted the possibilities of defining families of tilings by reference to generating triangles.

Two basic Möbius triangles gave us (4 2 4) and (3 2 6), the regular tilings and their associated families of semi-regular tilings. A Schwarz triangle consisting of paired Möbius triangles gave us (4 4/2 4), (3 6/2 3) and (6 3/2 6), three sub-families of double density tilings. Combining the last two of these Schwarz triangles gave us (3 6/3 6), a super-family of tilings all with density four.

We then looked at a Schwarz triangle composed of three (3 2 6) Möbius triangles and arrived at (6/2 2 6), the star tilings family of density three. This in its turn was paired up to produce (6/2 6/2 6/2) and (6 6/4 6), two further sub-families but with density six. Finally these two were combined to produce (6/2 6/3 6) the twelvefold super-family of tilings.

One further possibility presented itself within our grids of Möbius triangles, the (3 3 3) family, but was found to offer nothing not already seen.

Coxeter, in his *Regular Polytopes*, admits only four possible tiling families: (4 2 4), (3 2 6), (6 3/2 6) and (3 3 3), those that do not contain any compound polygons.

Interestingly these include one of each shape of the possible generating triangles that can be reflected and rotated in the plane. These have angles (45°, 90°, 45°), (60°, 90°, 30°), (30°, 120°, 30°) and (60°, 60°, 60°) respectively and are shown on the left in Figure 79. The seven other compound generating triangles we have explored are shown on the right.

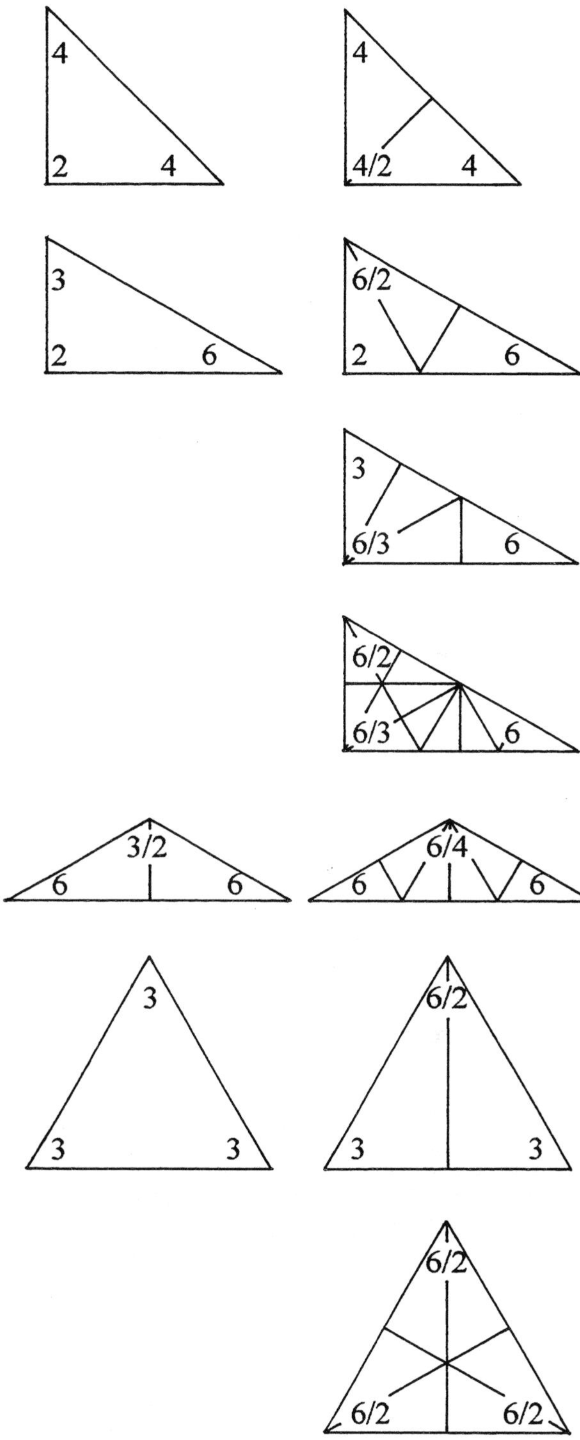

Fig.79: Generating Triangles

SO WHAT FOLLOWS?

In the foregoing we have seen that where compound polygons have been included, the tilings have not run riot into an infinity of possibilities as might be feared.

They have in fact produced a further seven families of tilings beyond the four 'accepted' ones, which have proved as interesting and as structured within their respective families as those original four.

All that is needed to accept these compound tilings into the realms of uniformity is to accept the compound polygons from which they are generated in their own right as equal members of a complete? set of possible polygons. This was proposed in the previous volume of this series and the consistency of the results produced in the present volume appear to justify the proposal.

Our next task will be to look at the polyhedra, where either fewer or lesser valued polygons meet at any one vertex, so that the figure produced folds up out of the two dimensions of our plane and wraps around to enclose a portion of three dimensional space.

Here we will again include compounds with some surprising results, allowing the classification of some 'not irregular' solids, such as Kepler's 'Stella Octangula', that were previously relegated to the miscellany outside the 'uniform polyhedra'.

Following on from this, the polyhedra themselves will be assembled in aggregate as the closepacks filling three dimensional space, which in their turn, by having fewer or lesser valued polyhedra around an edge, can be folded into four dimensions as the hypersolids.

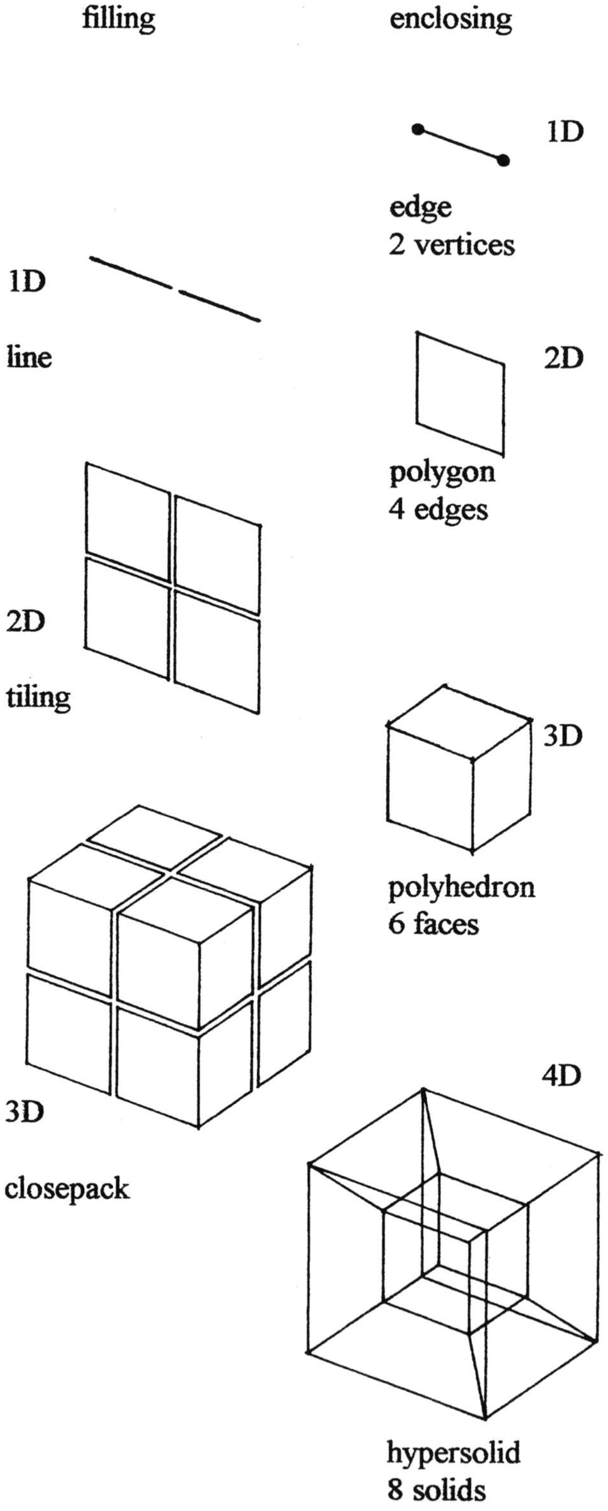

Fig.80: So What Follows?

BIBLIOGRAPHY

Abbott, E A 1962
Flatland
Blackwell

Coxeter, Longuet-Higgins & Miller 1953
Uniform Polyhedra
Phil. Trans. R. Soc. Lond. A246, p401

Coxeter, H S M 1963
Regular Polytopes
Macmillan

Critchlow, K 1969
Order in Space
Thames & Hudson

Grünbaum, B & Shephard, G C 1989
Tilings and Patterns: An Introduction
Freeman

Hargittai, I & M 1994
Symmetry
Shelter

Taylor, P 1990
The Geometry Trip
Nattygrafix

Taylor, P 1995
Additions to the Uniform Polyhedra
Nattygrafix

Taylor, P 1997
The Complete? Polygon
Nattygrafix

The Open University 1994
Tilings
M336 Block 1 Unit IB1

Weyl, H 1952
Symmetry
Princeton UP

ALSO PUBLISHED BY NATTYGRAFIX:

ADDITIONS TO THE UNIFORM POLYHEDRA by Patrick Taylor 1995

'does provide some new geometrical insights'

'extends the concept of a uniform polyhedron in an entertaining and significant manner'

presents some new configurations for inclusion within the uniform polyhedra

40pp A4 paperback including
37 line drawings and 3 tables

ISBN 0 9516701 1 5 £5.00

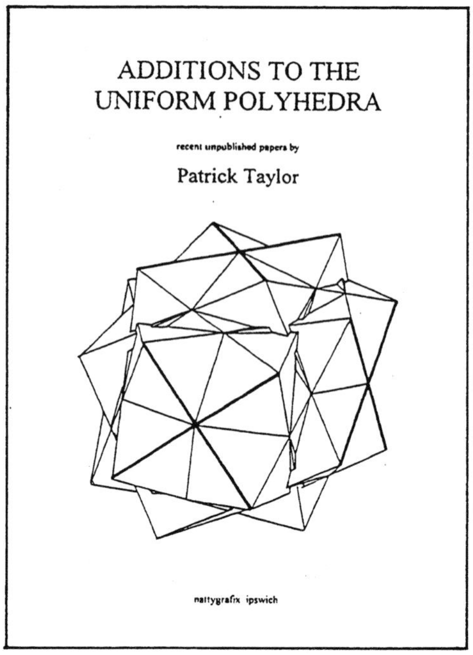

THE COMPLETE? POLYGON
by Patrick Taylor 1997

being the first part of several comprising the complete? polyhedra

presents the polygons as building blocks for the tilings, polyhedra etc.; including their generation from a simple vector model along with their truncations, quasitruncations and inverses; regular, star, compound, cross and double polygons; stellations and inscriptions.

60pp A4 paperback including
28 line drawings and 21 tables

ISBN 0 9516701 2 3 £6.00

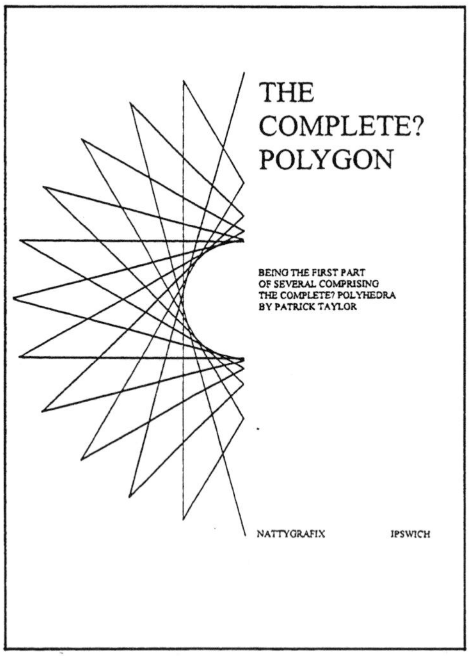